Homework Book

MyMaths
for Key Stage 3

1c

Powered by **MyMaths**.co.uk

OXFORD
UNIVERSITY PRESS

OXFORD
UNIVERSITY PRESS

Great Clarendon Street, Oxford OX2 6DP

Oxford University Press is a department of the University of Oxford.
It furthers the University's objective of excellence in research, scholarship,
and education by publishing worldwide in

Oxford   New York

Auckland   Cape Town   Dar es Salaam   Hong Kong   Karachi
Kuala Lumpur   Madrid   Melbourne   Mexico City   Nairobi
New Delhi   Shanghai   Taipei   Toronto

With offices in

Argentina   Austria   Brazil   Chile   Czech Republic   France   Greece
Guatemala   Hungary   Italy   Japan   Poland   Portugal   Singapore
South Korea   Switzerland   Thailand   Turkey   Ukraine   Vietnam

© Oxford University Press

The moral rights of the authors have been asserted

Database right Oxford University Press (maker)

First published 2013

British Library Cataloguing in Publication Data

Data available

ISBN 978-0-19-830446-3
10 9 8 7 6 5 4 3

Printed in Great Britain by Ashford Colour Press Ltd., Gosport

Paper used in the production of this book is a natural, recyclable product made from wood grown
in sustainable forests. The manufacturing process conforms to the environmental regulations to
the country of origin.

Contents

Jane's children each grew a sunflower. These values are the heights that they recorded on a particular day in June.

Alex 1.14 m   Jamie 140 cm   Matty 0.94 m   Andrew 1 m 4 cm

Arrange these values in order of size, smallest first, and write whose sunflower grew the tallest.

- -

The values in metres are 1.14, 1.4, 0.94 and 1.04.
In order these are 0.94, 1.04, 1.14 and 1.4 so Jamie had the tallest sunflower.

1 Write the value of these digits in the number 251.087
   a 1   **b** 2   **c** 8   **d** 0

2 The number 3.45 stands for 3 units, 4 tenths and 5 hundredths.
   What do these numbers stand for?
   a 4.21   **b** 25.1   **c** 12.037   **d** 138.0604

3 Write each of these as a decimal number.
   a one ten, seven units and five hundredths
   b nine units, five tenths, one hundredth and three thousandths
   c one hundred, two tens, four units and seven thousandths
   d eight tenths and four ten-thousandths.

4 Increase these numbers by seven tenths.
   a 3   **b** 0.28   **c** 16.43   **d** 0.998

5 Decrease these numbers by five hundredths.
   a 6.48   **b** 2.35   **c** 12.64   **d** 0.205

6 Write each list of numbers in order, starting with the smallest.
   a 8.4   8.3   8.28   8.35   8
   b 7.8   7.86   7.81   7.08   7.818
   c 5.92   9.52   5.29   52.9   0.529
   d 0.42   1.4   1.042   1.204   0.142
   e 0.003   ·0.03   0.036   0.603   0.063

Example

a	Multiply 0.36 by	**i**	10	**ii**	100	**iii**	1000
b	Divide 24 by	**i**	10	**ii**	100	**iii**	1000

You can use patterns to work out the answers to questions involving multiplying and dividing by 10, 100 and 1000.

a **i** $0.36 \times 10 = 3.6$ **b** **i** $24 \div 10 = 2.4$

  ii $0.36 \times 100 = 36$ **ii** $24 \div 100 = 0.24$

  iii $0.36 \times 1000 = 360$ **iii** $24 \div 1000 = 0.024$

1 Calculate

  a 34×10 **b** 8×1000

  c 12×100 **d** 4.1×10

  e 2.04×1000 **f** 45.2×100

  g 9.11×10 **h** 5.314×100

2 Calculate

  a $240 \div 10$ **b** $3500 \div 1000$

  c $650 \div 100$ **d** $4.1 \div 10$

  e $456 \div 1000$ **f** $992 \div 100$

  g $12 \div 10$ **h** $25 \div 100$

3   (×10) (×1000) (×100) (÷100) (÷10) (÷1000)

Write which of these operations must be applied to 3.056 to give

  a   305.6   **b**   0.3056

  c   3056   **d**   0.003 056

4 Write the operation that is needed to change

  a   2.56 to 25.6   **b**   56.12 to 0.5612

  c   0.35 to 0.000 35   **d**   314.3 to 0.3143

  e   1.052 to 105.2   **f**   15 to 0.15

  g   214.45 to 21 445   **h**   0.010 02 to 10.02

1c Negative numbers

Calculate	a $-25 - -10$	b $45 \div -9$	c $(-15)^2$
	a $-25 - -10$	b $45 \div -9$	c $(-15)^2$
	$= -25 + 10$	$= -5$	$= -15 \times -15$
	$= -15$		$= 225$

1 Put each of these sets of numbers in ascending order
(lowest to highest).

a -2 -3 4 0 -7

b 12 -16 -23 31 -50

c -2 -6.1 -1.8 -6.4 -2.5

2 Insert a < or > symbol between each pair of numbers to
make a true statement.

a $0 \square -3$ b $-6 \square 8$ c $-4 \square -7$

3 Write the difference between these temperatures.

a $-3\,°C$ and $0\,°C$ b $-1\,°C$ and $8\,°C$

c $3\,°C$ and $-5\,°C$ d $-12\,°C$ and $-7\,°C$

e $-15\,°C$ and $15\,°C$ f $-18\,°C$ and $-21\,°C$

4 Calculate

a $7 - 10$ b $-5 + 6$ c $-12 + 4$

d $-4 - 2$ e $-9 + 9$ f $20 - 40$

5 Calculate

a $8 + -3$ b $10 - 4$ c $-5 + 2$

d $6 - -9$ e $-12 + -8$ f $-15 - -5$

6 Find the missing numbers.

a $-2 \times -3 = \square$ b $12 \times \square = -24$ c $\square \div 3 = -5$

d $-6 \times \square = -30$ e $16 \div -2 = \square$ f $\square \div -50 = 2$

g $(-4)^2 = \square$ h $18 \times \square = -9$ i $(9 \div \square)^3 = -27$

1d Mental addition and subtraction

Example

a Use shopkeeper's subtraction to calculate $683 - 385$
b Use compensation to calculate $41.2 - 14.9$

- -

a Begin at 385 and count up to 683 in chunks.
$385 + \mathbf{15} = 400$ $400 + \mathbf{280} = 680$ $680 + \mathbf{3} = 683$
So, $683 - 385 = \mathbf{15} + \mathbf{280} + \mathbf{3} = 298$

b Subtract the nearest whole number and then compensate.
$41.2 - 14.9 = 41.2 - 15.0 + 0.1$ Subtract 15 and add 0.1 back on.
$= 26.3$

1 Calculate these using a mental method.
  a $32 + 18$ **b** $56 + 44$ **c** $84 - 14$ **d** $2.6 + 7.4$
  e $19.2 - 7.2$ **f** $12.4 + 7.6$ **g** $1.75 + 3.25$ **h** $11.75 - 6.25$

2 Use partitioning to calculate these mentally.
  a $4.1 + 5.8$ **b** $17.4 + 12.4$ **c** $24 - 14.6$ **d** $13.9 - 5.6$
  e $24.3 + 16.9$ **f** $8.3 - 5.8$ **g** $35.7 + 22.8$ **h** $45.4 - 17.8$

3 Use compensation to calculate these mentally.
  a $8.4 + 6.9$ **b** $44.5 + 12.8$
  c $16 - 11.9$ **d** $45 - 13.95$
  e $29.35 + 17.95$ **f** $12.55 - 6.85$
  g $14.73 + 9.85$ **h** $65.43 - 28.95$

4 Use 'shopkeeper's subtraction' to calculate these mentally.
  a $512 - 295$ **b** $934 - 475$ **c** $3264 - 1294$
  d $5276 - 2689$ **e** $8821 - 6543$ **f** $9549 - 4363$

5 Use doubling and halving to calculate these mentally.
  a 8×25 **b** 23×50 **c** 12×15
  d 19×25 **e** 9×45 **f** 32×15

6 Use compensation to calculate these mentally.
  a 32×19 **b** 21×39 **c** 45×21

MyMaths.co.uk
Q 1380   SEARCH

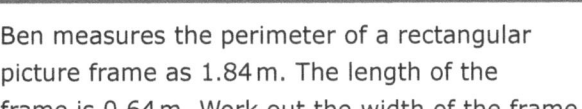

Example

Ben measures the perimeter of a rectangular picture frame as 1.84 m. The length of the frame is 0.64 m. Work out the width of the frame.

0.64 m

Two lengths are

$$0.64$$
$$+ \ 0.64$$
$$\overline{1.28 \,\text{m}}$$

$$1.\overset{7}{8}\overset{1}{4}$$
$$- \ 1.28$$
$$\overline{0.56 \,\text{m}}$$

Two widths are 0.56 m so one width is $\frac{1}{2}$ of 0.56 m = 0.28 m.

---

1 Calculate these using a written method.

  a 6.2 + 8.4   **b** 9.73 + 3.15
  c 3.8 + 2.9   **d** 5.36 + 4.94
  e 1.2 + 7   **f** 3.05 + 8.3
  g 2.73 + 1.4   **h** 7.27 + 1.845
  i 12 + 4.96 + 0.375 **j** 0.455 + 32 + 16.29

2 Calculate these using a written method.

  a 24.1 − 11   **b** 43.7 − 22.4
  c 31.2 − 25.4   **d** 1.25 − 0.96
  e 4.73 − 1.9   **f** 75.63 − 24.8
  g 15 − 5.2   **h** 6.8 − 3.39
  i 21 − 7.26   **j** 42.1 − 5.328

3 Use a written method to solve these problems.

  a In the United Kingdom, life expectancy at birth is 81 years for girls and 76.6 years for boys. How much longer can a girl expect to live than a boy?

  b In 2005, the population of the United Kingdom was estimated as 60.2 million people, of which 50.43 million lived in England. How many people lived in other areas of the United Kingdom?

  c Kirsten and Steve have £40 to spend on a two-course meal. For the main course, Kirsten chooses the lamb at £15.95 and Steve chooses the salmon at £11.49. How much money do they have to spend on their desserts?

 **MyMaths**.co.uk

🔍 1007, 1381   **SEARCH**

Number Whole numbers and decimals

Example

Change this time in hours to a time in hours and minutes.

| 5.2 |

0.2 hours = 0.2 × 60 minutes = 12 minutes
So 5.2 hours is 5 hours 12 minutes.

1 Remembering that, for example, 14.2 on a calculator means £14.20 when dealing with money, work these out.
   a The cost of 8 loaves of bread at £1.05 each.
   b The amount obtained by each person when a lottery win of £150 is shared equally between a syndicate of 12 people.
   c The cost of a class set of 30 pencils at 12p each.

2 Remembering that, for example, 28.4 on a calculator means 28 m and 40 cm when dealing with length, work these out.
   a The length in m and cm of five Ford Focus cars parked nose to tail if each car is 4.34 m in length.
   b The total length in m and cm of three friends laid head to foot on the ground if they measure 1.58 m, 180 cm and 1 m 65 cm.
   c The thickness in cm of a mathematics textbook if a pile of 30 textbooks has a height of 0.75 m.

3 These calculator displays give times in hours.
   Change them to hours and minutes.
   a
   | 7.25 |
   b
   | 2.1 |
   c
   | 4.333333333 |

4 Write a mental estimate for each of these calculations and then use a calculator to work out the exact answer.
   a $3.9 + 6.15 - 4.84$ b $31.35 - (8.18 \times 3.2)$ c $\frac{19.8}{4.4} + 3.61$
   d $(2.1)^3$ e -3.24×-1.96 f $-25.99 - -14.38$

MyMaths.co.uk

Q 1167, 1932 | SEARCH

Write the reading on this scale.

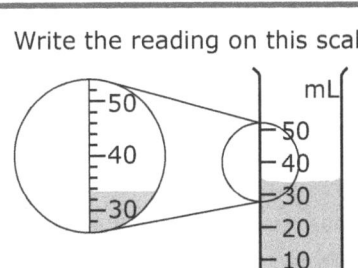

The level of the liquid in this measuring cylinder is between 30 and 40 ml. There are 5 divisions between 30 and 40 ml so each division is 10 ÷ 5 = 2 ml.
The reading is 34 ml of liquid.

1 Match the objects on the left with an appropriate measurement from the table on the right.

Length of pencil
Length of a book
Height of door
Capacity of a cup

2 m
17 cm
250 ml
25 cm

2 Write the readings on these scales.

a

b

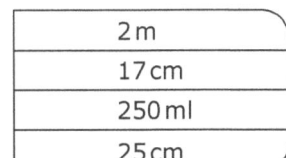

3 Estimate the readings on these scales.

a

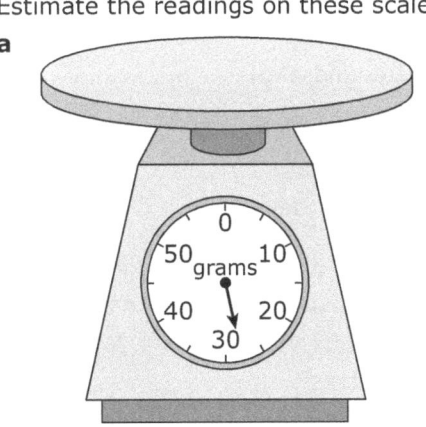

b

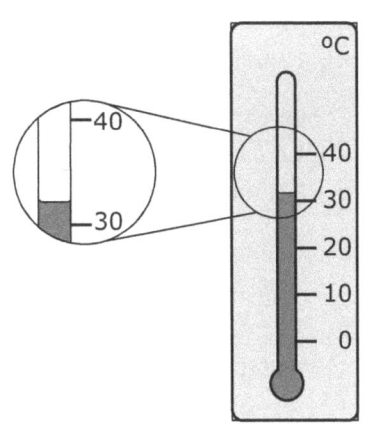

Example

Find the area of this rectangle in
a cm²   **b** mm²

1.5 cm

8 mm

a 8 mm ÷ 10 = 0.8 cm   Convert mm to cm
   1.5 cm × 0.8 cm = 1.2 cm²   Then use area = length × width
b 1.2 cm² = 1.2 × 100   Convert cm² to mm²
        = 120 mm²

---

1 Convert these metric measurements to the units indicated
   in brackets.
   a 12 m (cm)   **b** 24 cm (mm)
   c 75 cl (ml)   **d** 4.5 litres (ml)
   e 1.25 kg (g)   **f** 0.55 tonnes (kg)
   g 5 m² (cm²)   **h** 6.5 ha (m²)

2 Convert these metric measurements to the units indicated
   in brackets.
   a 430 cm (m)   **b** 75 mm (cm)
   c 75 cl (litres)   **d** 2750 ml (litres)
   e 815 g (kg)   **f** 9045 kg (tonnes)
   g 43 500 cm² (m²)   **h** 125 000 m² (ha)

3 Convert these imperial units to their rough metric equivalents.
   a 21 feet (m)   **b** 64 inches (cm)
   c 12 gallons (litres)   **d** 110 pounds (kg)

4 Convert these measurements of time to the units indicated in brackets.
   a 12 minutes (seconds)   **b** 7 days (hours)
   c 220 minutes (hours)   **d** 200 hours (days)

5 a A shortbread biscuit weighs 30 g. The biscuits are sold in packs
        of 8. What is the mass of a pack of biscuits in **i** grams **ii** kg?
   b The distance between Clutton and Tattenhall is 5.8 km. What is this
        distance in **i** metres **ii** millimetres?

MyMaths.co.uk

Q 1091, 1191 SEARCH

Example

Calculate the **a** perimeter **b** area of this rectangle.

9 cm

14 cm

a Perimeter = 14 cm + 9 cm + 14 cm + 9 cm

   = 46 cm

b Area = length × width

   = 14 cm × 9 cm

   = 126 cm²

1 Calculate the **i** perimeter **ii** area of each rectangle.

a 5 cm

9 cm

b 22 cm

14 cm

c 9.5 cm

12 cm

2 Calculate the missing lengths from each of these rectangles.

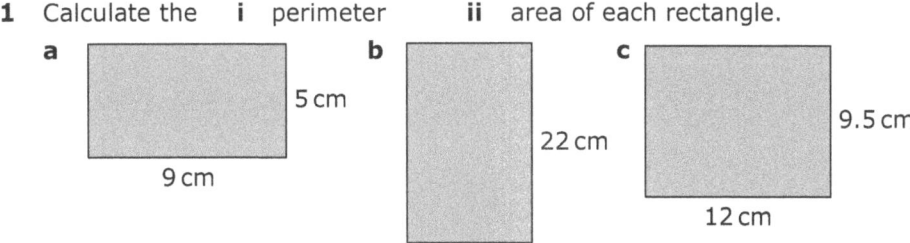

a Area = 81 cm² *h* cm

9 cm

b Area = 50 cm² 4 cm

p cm

c Area = 96 cm² *x* cm

12 cm

3 Calculate the perimeter and area of these shapes.

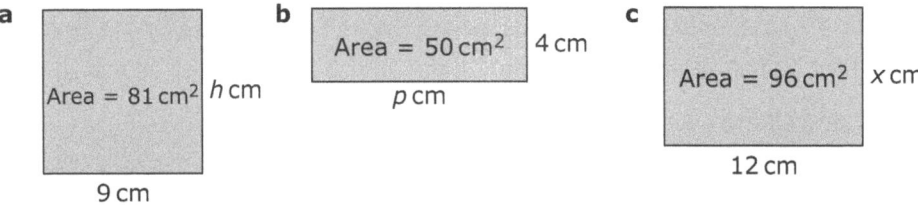

a 3 cm

5 cm

b 9 mm 2 mm

10 mm

6 mm

c 12 m

6 m

4 m

12 m

20 m

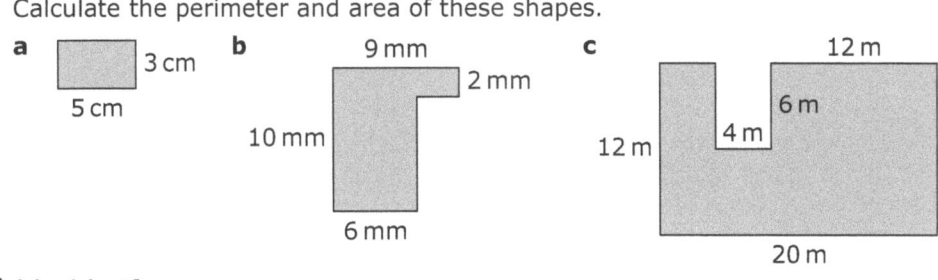

Example

Calculate the perpendicular height of this triangle.

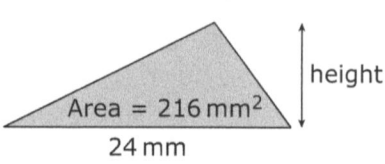

height

Area = 216 mm²

24 mm

Area $= \frac{1}{2}$(base × height)

$2 \times 216\,mm^2 = 24\,mm \times height$

$\frac{432}{24} = height$

So the height of the triangle is 18 mm.

1 Calculate the   **i** perimeter   **ii** area of each triangle.

a

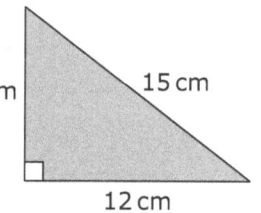

9 cm   15 cm

12 cm

b

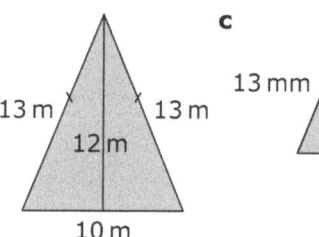

13 m   13 m

12 m

10 m

c

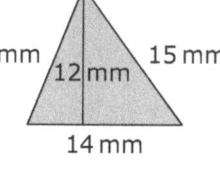

13 mm   15 mm

12 mm

14 mm

2 Calculate the missing lengths on each of these triangles.

a

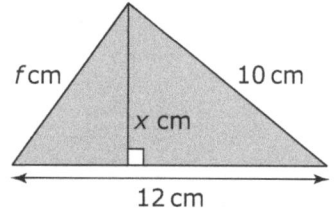

f cm   10 cm

x cm

12 cm

Perimeter = 27 cm
    Area = 24 cm²

b

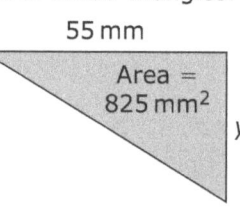

55 mm

Area = 825 mm²

y mm

3 Calculate the area of these shapes.

a

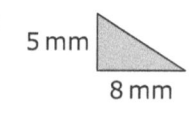

5 mm

8 mm

b

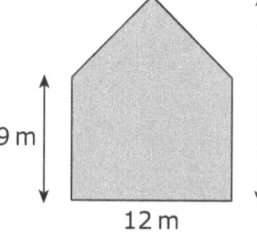

9 m

15 m

12 m

c

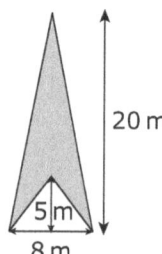

20 m

5 m

8 m

MyMaths.co.uk

Q 1129   SEARCH

Example

Calculate the shaded area of this shape.

20 mm

10 mm

8 mm

The outer shape is a rectangle so its area is 20 × 10 = 200 mm²

The unshaded area is in the shape of a trapezium its area is $\frac{1}{2}$(20 + 8) × 10 = 140 mm²

So the shaded area is 200 − 140 = 60 mm²

1 Calculate the area of each shape.

a
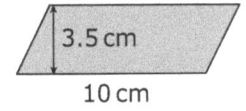
3.5 cm
10 cm

b
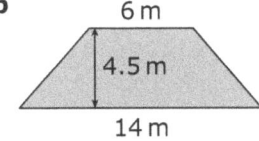
6 m
4.5 m
14 m

c
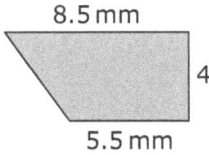
8.5 mm
4 mm
5.5 mm

2 Calculate the area of each shape.

a
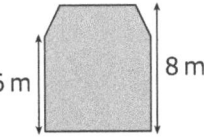
5 m
6 m
8 m
7 m

b
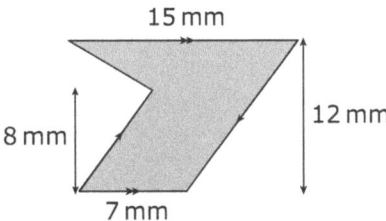
15 mm
8 mm
12 mm
7 mm

3 Calculate the missing lengths on these shapes.

a
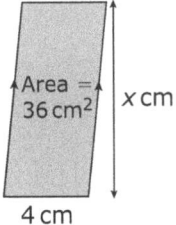
Area = 36 cm²
x cm
4 cm

b
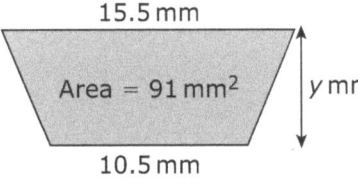
15.5 mm
Area = 91 mm²
y mm
10.5 mm

4 A path crosses a garden as shown in the diagram. Calculate the area of the path.

12 m

8 m

10 m

Example

A cube has a surface area of 541.5 cm². Calculate
a the area of one face
b the length of one edge.

- -

a   Area of one face = 541.5 cm² ÷ 6
                          = 90.25 cm²

b   Length of one side = $\sqrt{90.25}$
                           = 9.5 cm

1   Pair up the cuboids that have the same surface area.
Show all your working.

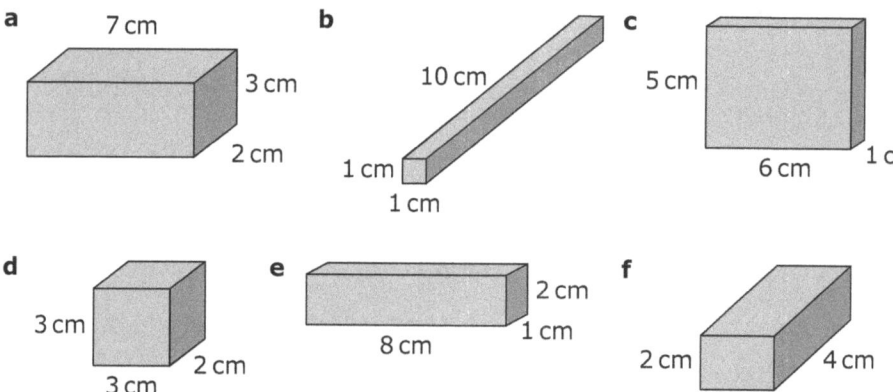

a   7 cm   3 cm   2 cm

b   10 cm   1 cm   1 cm

c   5 cm   6 cm   1 cm

d   3 cm   3 cm   2 cm

e   8 cm   2 cm   1 cm

f   2 cm   3 cm   4 cm

2   Calculate the area of one face of a cube with a surface
area of
a   96 cm²   **b**   37.5 m²   **c**   7350 mm²

3   Calculate the length of one edge of each cube in
question **2**.

4   Flora has a set of building bricks that measure 4 cm by
4 cm by 6 cm. How many of these bricks can she fit in a
box that has length 42 cm, width 16 cm and height 8 cm?

MyMaths.co.uk

Q 1107   SEARCH

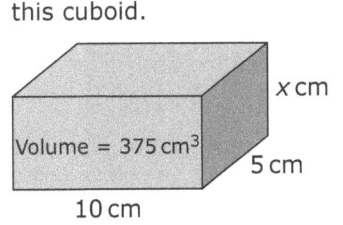

Calculate the height of this cuboid.

Volume = 375 cm³

x cm

5 cm

10 cm

Volume = length × width × height

$375\,\text{cm}^3 = 10\,\text{cm} \times 5\,\text{cm} \times x$

$x = \dfrac{375}{50}$

$= 7.5\,\text{cm}$

1 Pair up the cuboids that have the same volume.
Show all your working.

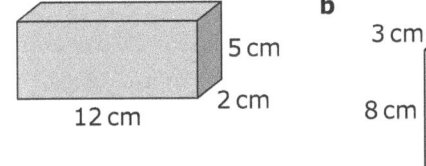

a   5 cm   2 cm   12 cm

b   3 cm   3 cm   8 cm

c   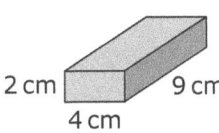   2 cm   9 cm   4 cm

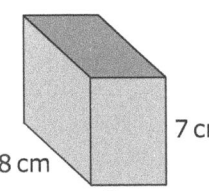

d   7 cm   8 cm   5 cm

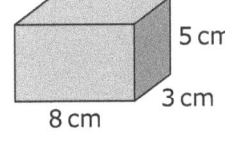

e   5 cm   3 cm   8 cm

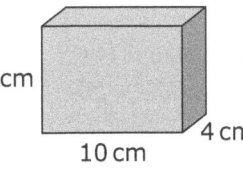

f   7 cm   10 cm   4 cm

2 Calculate the missing lengths on each cuboid.

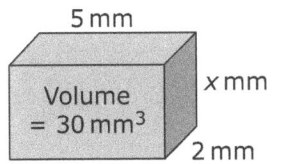

a   5 mm x mm   Volume = 30 mm³   2 mm

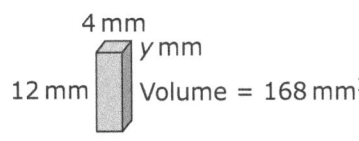

b   4 mm y mm   12 mm   Volume = 168 mm³

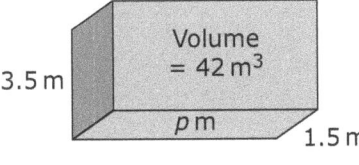

c   3.5 m   Volume = 42 m³ p m   1.5 m

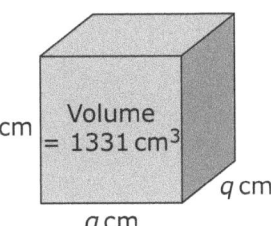

d q cm   Volume = 1331 cm³ q cm q cm

Example

Grace and Isobel both substitute $p = 5$ into the expression $2p^2$.
Grace gets 50 and Isobel gets 100. Who is correct?

Grace is correct as she has worked out $2 \times 5^2 = 2 \times 25 = 50$.
Isobel has worked out $2 \times 5 = 10$ and then squared her answer
to get 100. BIDMAS reminds us to calculate indices first.

1 Write an algebraic expression for each of these.
  a Jason has x pounds in his bank account. He deposits £5.
    How much money is in his account now?
  b There are 12 birds sitting on a streetlight. y birds fly away.
    How many birds are there on the streetlight now?
  c Victoria puts k cakes onto each of her three friends' plates.
    She then adds another cake to each plate. How many cakes
    has she used?
  d Alexander paints m pictures. In each picture there are n trees.
    How many trees has he painted?

2 Given that $x = 2$ and $y = 8$, find the value of these expressions.
  a $x + 10$ **b** $y - 3$ **c** $2y$
  d $\dfrac{y}{x}$ **e** $2y + 3$ **f** xy
  g $3(x + 1)$ **h** $x^2 + y$

3 Match one of these number cards with each of these expressions,
  given that $a = -2$.

  | 4 | −6 | 8 | −8 | 1 | 6 | −4 | −1 |

  a a^2 **b** $2a + 5$ **c** $2a$
  d $\dfrac{a}{2}$ **e** $5a + 2$ **f** $a^3 + 2$
  g $2(a + 5)$ **h** $2a^2$

Example

Which of these expressions is equivalent to $3x^2 - x - 6 - x + 6$?

$3x^2 - 12$ x^2 $3x^2 - 2x$

$$3x^2 - x - 6 - x + 6 = 3x^2 - x - x + 6 - 6$$
$$= 3x^2 - 2x$$

1 Simplify these expressions.

   a $x + x + x + x$ **b** $3a + 4b$

   c $2 \times y$ **d** $a \times a$

   e $9k - 4k$ **f** $8p + 3p - 5p$

   g $4m - 2m + 3$ **h** $12b + 3c - 4b + 5c$

   i $5p - 4q + p$ **j** $2t + 8t^2 - 2t$

   k $4ab + 8ab - 3ba$ **l** $\frac{1}{2}(6n - 4n^3)$

2 Sort these cards into pairs of equivalent algebraic expressions.
Show all your working.

$3p + 3p + 4p$ $pq + 7$ $3pq + 4qp$

$q - q + p - 7q$ $7 \times q \times p$

$5p + 2q + 5p - 2q$ $-7q + p$ $4pq - 3pq + 3 + 4$

3 Show that the perimeters of these shapes are equal.

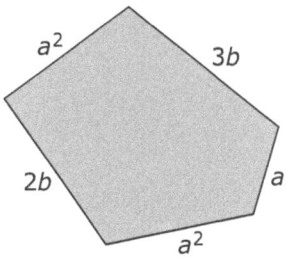
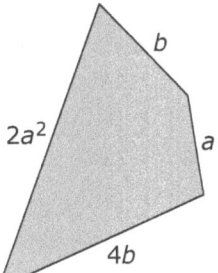

Example

Expand and simplify this algebraic expression.

$3(x + 5) - 2(x - 3)$

- -

$$3(x + 5) - 2(x - 3) = 3x + 15 - 2x + 6$$
$$= 3x - 2x + 15 + 6$$
$$= x + 21$$

1 Expand these brackets.

a	$2(x + 4)$		**b**	$3(y + 2)$
c	$5(p + 10)$		**d**	$4(q - 1)$
e	$a(b + 5)$		**f**	$x(y - 8)$
g	$m(12 - n)$		**h**	$a(a + 7)$
i	$2x(x + y)$		**j**	$3p(6 - 2p)$

2 Expand these brackets and simplify where possible.

a	$2(x + y + 3)$		**b**	$-4(a + 2b + c)$
c	$x(x + 3y - 1)$		**d**	$4(a + 1) + 2(a + 3)$
e	$3(q + 6) + 4(q - 3)$		**f**	$5(2x + 1) + 3(3x - 4)$
g	$2(t + 3) - 4(t + 1)$		**h**	$a(a + b) + b(a - 2b)$

3 Write an expression for
   a the perimeter
   b the area of this rectangle.

$3a - 2$

4

4 Lily thought of a number, subtracted 2 and then multiplied
   by 5. Gemma thought of a number, multiplied it by
   5 and then subtracted 10. Both girls got 20.
   a Find the number that Lily thought of.
   b Find the number that Gemma thought of.
   c What do you notice about these numbers?

   Will this always be true? If so, prove your answer using algebra.
   Write an algebraic expression and expand the brackets.

Convert £120 to Australian dollars using the formula $D = 2.5P$
where D is the number of Australian dollars and P is the
number of pounds.

$D = 2.5 \times 120$ Substitute 120 for P.

$= 300$

So £120 is 300 Australian dollars.

1 The perimeter of a square can be found using the formula $P = 4l$
where P is the perimeter and l is the length.

a Find P when $l = 6\,cm$.

b Find the perimeter of a square with sides $8\,mm$.

c Explain why this formula works.

2 The mass (m) of an object is measured in kilograms and cannot change.
The weight (w) of an object is measured in newtons (N) and depends
on gravity (g). Gravity is measured in N/kg (newtons per kg). On Earth
gravity is $10\,N/kg$ and on Mars gravity is $4\,N/kg$. The weight of an object
is found by multiplying its mass by gravity: $w = mg$.
A person has a mass of $50\,kg$. Work out their weight

a on Earth

b on Mars.

3 A plumber calculates the cost to his clients using the formula
$C = 45 + 9.5h$ where C is the cost in pounds and h is the number of
hours he takes to fix a problem.

a Work out the cost of a job that lasts 2 hours.

b What is the plumber's call-out fee? Explain your answer.

Example

Write a formula for the area of this parallelogram, expanding any brackets.

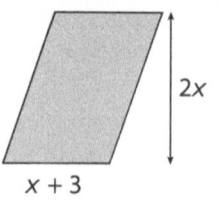

$2x$

$x + 3$

Area of a parallelogram
$$= \text{length} \times \text{height}$$
$$= (x + 3) \times 2x$$
$$= 2x(x + 3)$$
$$= 2x^2 + 6x$$

1 a How many edges are there on a cube? On five cubes?

  b Write a formula to find E, the number of edges on x cubes.

2 Philip is given £10 pocket money per week and £2.50 for each household chore that he completes.

  a Write a formula for M, the amount of money that he receives per week based on c, the number of chores that he completes.

  b Use your formula to work out how much money Philip receives if he washes his Dad's car and mows the lawn in one week.

3 The cost of spending 7 nights in a four-star hotel in Sorrento, Italy, is £560 per adult and £350 per child.

  a Write a formula for C, the total cost for a adults and c children taking this holiday.

  b Use your formula to work out how much this holiday would cost a family of 2 adults and 3 children.

4 One of the parallel edges of this trapezium is 3 times the length of the other.
Write a formula for the area of this trapezium.

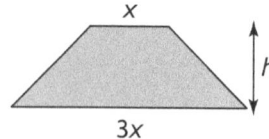

x

h

$3x$

Substitute x and $3x$ into the formula for the area of a trapezium and simplify.

a Write an expression for the perimeter of this rectangle.

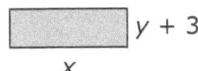

$y + 3$

x

b Simplify the expression by collecting like terms.

c Find the perimeter of the rectangle if $x = 5$ and $y = -1$.

- -

a $x + x + y + 3 + y + 3$

b $2x + 2y + 6$

c $(2 \times 5) + (2 \times -1) + 6 = 10 - 2 + 6$
$$= 14 \text{ units}$$

1 If $w = 5$, $x = 2$ and $y = -3$, find the value of each expression.

a $y + 5$ **b** $x - 9$ **c** $3w$ **d** x^2

e $2x + w$ **f** $3y + 5x$ **g** xy **h** $wx + 4$

i $2w^2$ **j** y^3 **k** $xy + w$ **l** $10 - wx$

2 Simplify these expressions by collecting like terms. Then find the value of each expression given that $a = -3$ and $b = 2$.

a $5b + 2b$ **b** $2a + 6b - a$ **c** $4b - 3a + b - 5$

d $2ab + 3ba$ **e** $2a^2 + 5a + b + a$ **f** $b^2 + 7b^2 + 10a$

3 **a** Write an expression for the area of this rectangle.

b Simplify your expression, expanding any brackets.

c Given that $n = 8$, work out the area of this rectangle.

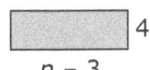

4

$n - 3$

4 Match the pairs of cards with equivalent expressions. Show your work.

$4x - 8$	$2(2x - 1)$	$4x - 2$	$4x^2 - x$
$x(4x - 1)$	$4(x - 2)$	$4x - 6$	$2(2x - 3)$

Example

Simplify these expressions.

a $3x \times 4$ **b** $3p \times 5q$ **c** $2a \times 6a$

---

a $3x \times 4 = 3 \times 4 \times x$
$\qquad\qquad = 12x$

b $3p \times 5q = 3 \times 5 \times p \times q$
$\qquad\qquad\quad = 15pq$

c $2a \times 6a = 2 \times 6 \times a \times a$
$\qquad\qquad\quad = 12a^2$

1 Simplify these expressions as fully as possible.

  a $a \times 3$ **b** $6p \times 2$ **c** $k \times 4k$ **d** $3p \times q$

  e $2m \times 5n$ **f** $4a \times 3b \times 2c$ **g** $2w \times 8xy$ **h** $5g \times 3g^2$

2 Expand and simplify these expressions.

  a $5(a + 3)$ **b** $4(x - 1)$

  c $10(2 - p)$ **d** $x(y + 3)$

  e $2(k + 3) + 4(k + 1)$ **f** $7(t + 2) - 4(t + 1)$

  g $a(a + 5) + a(a - 2)$ **h** $3p(5p - 2)$

  i $5x(3x + 2y)$ **j** $2a(3a + 4) + 6a(2a + 1)$

3 Write a simplified expression for the area of this rectangle.

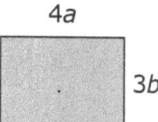

4 **a** Write a simplified expression for the area of this square.

   b If $x = 3$, find the area of this square.

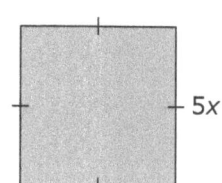

MyMaths.co.uk

Q 1178   SEARCH

Example

Simplify

a $\dfrac{15ab}{3b}$ **b** $\dfrac{a^2 + 4a}{a}$

- -

a $\dfrac{15ab}{3b} = \dfrac{{}^5\cancel{15} \times a \times \cancel{b}}{{}_1\cancel{3} \times \cancel{b}} = 5a$

b a is a factor of both a^2 and $4a$ and so it can be cancelled.

$\dfrac{a^2 + 4\cancel{a}}{\cancel{a}} = a + 4$

1 Simplify these expressions as fully as possible.

a $x \div 5$ **b** $2 \div y$ **c** $8m \div 2$ **d** $5k \div 10$

e $\dfrac{12a}{4}$ **f** $\dfrac{21b}{7b}$ **g** $\dfrac{3xy}{9x}$ **h** $\dfrac{24t^2}{3t}$

2 Mark Kim's homework. If you find a mistake then please correct it for her, explaining where she went wrong.

a $\dfrac{5x}{5} = x$ **b** $\dfrac{k + 7}{7} = k$ **c** $\dfrac{3d^2}{d} = 3d$

d $\dfrac{2ab^2}{4b} = 2ab$ **e** $\dfrac{10t + 18}{5} = 2t + 18$

3 Simplify these expressions.

a $\dfrac{10x + 4}{2}$ **b** $\dfrac{4a - 12}{4}$ **c** $\dfrac{10p - 15q}{5}$

d $\dfrac{18m + 12n}{6}$ **e** $\dfrac{x^2 + 3x}{x}$ **f** $\dfrac{5a + a^2}{a}$

g $\dfrac{b^2 - 7b}{b}$ **h** $\dfrac{8t - 4t^2}{2t}$ **i** $\dfrac{12f + 6f^2}{3f}$

4a Fraction notation

Write the angle 135°
as a fraction of a turn.

One complete turn is 360° so

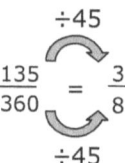

$\div 45$

$$\frac{135}{360} = \frac{3}{8}$$

$\div 45$

1   Giving your answer in its simplest form, write
    the fraction of this rectangle that is shaded
    a black   **b** grey   **c** white.

2   Copy this rectangle and shade these
    fractions in the colour indicated.
    a $\frac{1}{3}$ in red **b** $\frac{1}{6}$ in blue **c** $\frac{2}{5}$ in yellow.

    d Write the fraction that is unshaded.

3   Copy and complete these fraction equivalents.
    a $\frac{2}{5} = \frac{\square}{15}$ **b** $\frac{4}{9} = \frac{20}{\square}$ **c** $\frac{\square}{7} = \frac{6}{14}$ **d** $\frac{7}{\square} = \frac{56}{64}$

4   Write these angles as fractions of a turn. Give your
    answers in their simplest form.
    a 90°   **b** 270°   **c** 315°   **d** 225°   **e** 22.5°

5   The table shows the number of letters in the first 50 words
    in the children's book *Matilda* by Roald Dahl.

Word length	1	2	3	4	5	6	7	8	9	10
Number of words	1	8	7	9	10	3	7	1	2	2

    a Write the fraction of the words that are 4 letters in length.
    b Write the fraction of the words that are not 5 letters in length.
    c How many letters are there in the words that occur $\frac{4}{25}$
          of the time?

MyMaths.co.uk

Q 1062   **SEARCH**

4b Adding and subtracting fractions

Example

Calculate

a $\dfrac{3}{8} + \dfrac{2}{5}$ **b** $4\dfrac{3}{4} - 1\dfrac{1}{3}$

- -

a $\dfrac{3}{8} + \dfrac{2}{5} = \dfrac{15}{40} + \dfrac{16}{40}$ **b** $4\dfrac{3}{4} - 1\dfrac{1}{3} = \dfrac{19}{4} - \dfrac{4}{3}$

$\qquad = \dfrac{31}{40}$ $\qquad = \dfrac{57}{12} - \dfrac{16}{12}$

$\qquad\qquad\qquad\qquad = \dfrac{41}{12}$

$\qquad\qquad\qquad\qquad = 3\dfrac{5}{12}$

1 Find the missing numbers in these sets of equivalent fractions.

a $\dfrac{1}{4} = \dfrac{\square}{8} = \dfrac{5}{\square} = \dfrac{\square}{24} = \dfrac{15}{\square}$ **b** $\dfrac{3}{5} = \dfrac{\square}{15} = \dfrac{12}{\square} = \dfrac{\square}{40} = \dfrac{30}{\square}$

c $\dfrac{3}{7} = \dfrac{\square}{14} = \dfrac{15}{\square} = \dfrac{\square}{70} = \dfrac{60}{\square}$ **d** $\dfrac{7}{9} = \dfrac{\square}{18} = \dfrac{28}{\square} = \dfrac{\square}{81} = \dfrac{700}{\square}$

2 Change these improper fractions to mixed numbers.

a $\dfrac{7}{3}$ **b** $\dfrac{9}{2}$ **c** $\dfrac{11}{5}$ **d** $\dfrac{15}{4}$ **e** $\dfrac{43}{8}$

3 Change these mixed numbers to improper fractions.

a $1\dfrac{2}{3}$ **b** $3\dfrac{1}{8}$ **c** $2\dfrac{3}{4}$ **d** $1\dfrac{7}{10}$ **e** $4\dfrac{5}{6}$

4 Calculate these, giving your answer as a fraction in its simplest form or a mixed number.

a $\dfrac{3}{5} + \dfrac{1}{5}$ **b** $\dfrac{8}{9} - \dfrac{2}{9}$ **c** $\dfrac{3}{4} + \dfrac{3}{4}$ **d** $1\dfrac{1}{6} - \dfrac{5}{6}$ **e** $4\dfrac{2}{7} - 3\dfrac{3}{7}$

5 Calculate these. You will need to change one or both fractions to equivalent fractions.

a $\dfrac{1}{4} + \dfrac{5}{8}$ **b** $\dfrac{5}{6} - \dfrac{1}{3}$ **c** $\dfrac{1}{2} + \dfrac{1}{3}$ **d** $\dfrac{2}{3} + \dfrac{3}{4}$

e $3\dfrac{1}{2} - 1\dfrac{3}{5}$ **f** $1\dfrac{1}{3} + 3\dfrac{1}{2}$ **g** $4\dfrac{3}{8} + 1\dfrac{3}{4}$ **h** $2\dfrac{2}{5} - 1\dfrac{1}{6}$

Example

Change these fractions to decimals.

a $\frac{11}{40}$

b $\frac{5}{8}$

- -

a

$\times 2.5$

$\frac{11}{40} = \frac{27.5}{100}$

$\times 2.5$

so $\frac{11}{40} = 0.275$

b $\frac{5}{8} = 8\overline{)5.000}$

$\begin{array}{r} 0.625 \\ \hline 48 \\ \hline 20 \\ 16 \\ \hline 40 \end{array}$

so $\frac{5}{8} = 0.625$

---

1 Copy and complete the table.

Fraction	$\frac{1}{2}$		$\frac{1}{10}$	$\frac{3}{4}$			$\frac{1}{3}$	
Decimal		0.25			0.01	0.2		0.6

2 Pair these cards if the two numbers are equivalent. Do not use a calculator and show your working.
Write the odd one out.

$\frac{7}{20}$ 0.9 0.8 $\frac{3}{25}$ 0.35

0.4 $\frac{4}{5}$ $\frac{9}{10}$ 0.12

3 Write each of these decimals as a fraction in its simplest form.

a 0.6 **b** 0.03 **c** 0.45 **d** 0.625 **e** 1.56

4 Write each of these fractions as a decimal without using a calculator.

a $\frac{2}{5}$ **b** $\frac{3}{8}$ **c** $\frac{11}{25}$ **d** $\frac{17}{40}$ **e** $\frac{4}{9}$

5 These are Michael's exam results. Write them in order, starting with the subject in which he scored the highest mark.

Chemistry $\frac{32}{40}$ English $\frac{39}{60}$ Physics $\frac{36}{50}$ Mathematics $\frac{49}{70}$

MyMaths.co.uk

1016 SEARCH

Work out $\frac{7}{9}$ of $\frac{2}{3}$ of 162.

$\frac{1}{3} \times 2 \times 162 = \frac{1 \times 324}{3} = \frac{324}{3} = 108$

$\frac{1}{9} \times 7 \times 108 = \frac{1 \times 756}{9} = \frac{756}{9} = 84$

1 Use a written method to calculate these.

a $\frac{2}{9}$ of £63   **b** $\frac{5}{12}$ of 120 m   **c** $\frac{3}{8}$ of 40 kg   **d** $7 \times \frac{1}{3}$

e $\frac{5}{6}$ of $54   **f** $\frac{3}{11}$ of 132 km   **g** $20 \times \frac{4}{9}$ **h** $\frac{2}{13} \times 104$

2 Isla bakes 36 fairy cakes. She covers $\frac{1}{4}$ of them in purple icing, $\frac{2}{9}$ of them in pink, 11 in pale blue and all the rest in chocolate except for one which is eaten by her mother.

a Calculate the number of cakes that have purple icing.

b Calculate the number of cakes that have pink icing.

c How many cakes are covered in chocolate?

3 Work out $\frac{2}{5}$ of $\frac{1}{4}$ of 180.

4 Carbon-14 has a half-life of 5370 years. This means that after 5370 years half of the atoms have decayed.
How many years will it take for a quantity of carbon-14 to become $\frac{1}{8}$ of its original size?

5 Calculate

a $4 \div \frac{1}{7}$ **b** $5 \div \frac{1}{9}$ **c** $7 \div \frac{2}{3}$ **d** $9 \div \frac{3}{5}$ **e** $12 \div \frac{8}{9}$

6 **a** Leaf fills plant pots with compost from a 30 litre bag.
Each plant pot will hold $\frac{3}{4}$ litre of compost.
How many pots can he fill?

b David makes plant labels by cutting $8\frac{1}{2}$ cm lengths from a plank of wood 153 cm long.
How many labels can he cut?

4e Percentages

1 Write these percentages as decimals.

   a 24%   **b** 19%   **c** 70%   **d** 7%   **e** 120%

2 Write these decimals as percentages.

   a 0.51   **b** 0.08   **c** 0.6   **d** 0.175   **e** 1.5

3 Write these percentages as fractions in their simplest form.

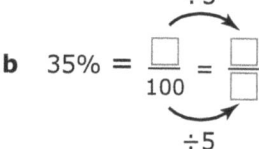

   a $20\% = \dfrac{\square}{100} = \dfrac{\square}{5}$ **b** $35\% = \dfrac{\square}{100} = \dfrac{\square}{\square}$

   c 74%   **d** 56%

4 Write these fractions as percentages using equivalent fractions.

   a $\dfrac{3}{4} = \dfrac{\square}{100} = \square\,\%$ **b** $\dfrac{4}{5} = \dfrac{\square}{100} = \square\,\%$

   c $\dfrac{3}{20}$ **d** $\dfrac{11}{25}$

5 Pair these cards if they show equivalent numbers.
   Which is the odd card out?

   40%   0.45   0.045   450%   4%

   0.4   4.5   4.5%   45%

MyMaths.co.uk

Q 1031   SEARCH

Example

> Change $\frac{1}{32}$ into **a** a decimal   **b** a percentage using a calculator.
>
> -
>
> **a** Type ⟨1⟩⟨÷⟩⟨3⟩⟨2⟩ into your calculator. This gives 0.031 25.
>
> **b** To change 0.031 25 into a percentage, type ⟨×⟩⟨1⟩⟨0⟩⟨0⟩ giving 3.125%.

1 Pair these cards if they show equivalent numbers. Which is the odd card out?

75%	1.75	0.175	7.5%	17.5%

175%	7.5	0.075	0.75

2 Change these fractions into percentages.

a $\frac{3}{4} = \frac{\square}{100} = \square\%$ **b** $\frac{2}{5} = \frac{\square}{100} = \square\%$ **c** $\frac{7}{20}$ **d** $\frac{18}{25}$

3 Use a calculator to change these fractions to
i decimals
ii percentages.
Give your answers to 2 decimal places where appropriate.

a $\frac{13}{20}$ **b** $\frac{7}{40}$ **c** $\frac{11}{16}$ **d** $\frac{4}{9}$ **e** $1\frac{1}{3}$

4 Change these decimals to fractions in their simplest form by first writing each one as a fraction out of 100.

a $0.26 = \frac{26}{100} = \frac{\square}{50}$ **b** $0.95 = \frac{\square}{100} = \frac{\square}{20}$

c 0.6   **d** 0.04

Example

Calculate the unknown angle.

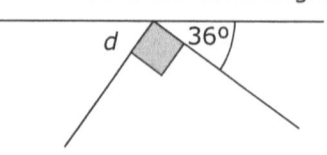

The known angles are 36° and 90°.
Angles on a straight line = 180°
$d + 36° + 90° = 180°$
$d = 180° - 36° - 90°$
$d = 54°$

1 Calculate the unknown angles.

a

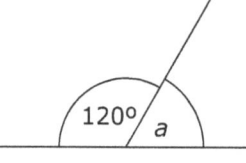

120° a

b

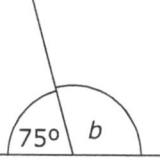

75° b

c

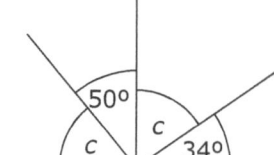

50°
c
c   34°

2 Calculate the unknown angles.

a

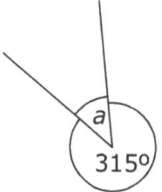

a
315°

b

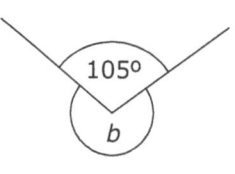

105°
b

c

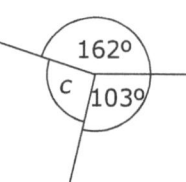

162°
c   103°

d

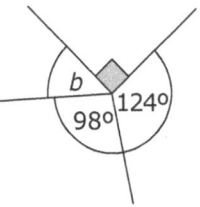

b
98° 124°

e

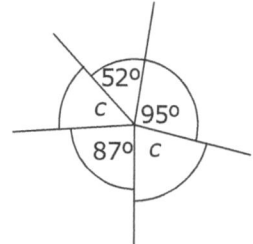

52°
c   95°
87° c

f

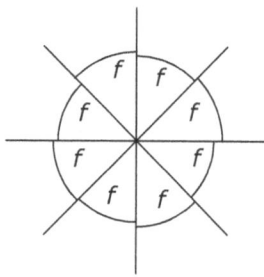

f   f
f   f
f   f
f   f

Find the angles of this triangle, giving reasons for your answers.

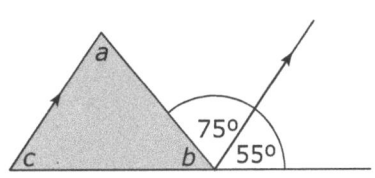

$a = 75°$ (alternate angles)
$b = 180° - 75° - 55° = 50°$
(angles on a straight line)
$c = 55°$ (corresponding angles)

1 Calculate the unknown angles, giving reasons for your answers.

a

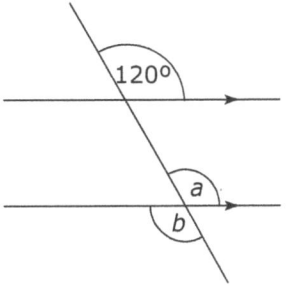

b

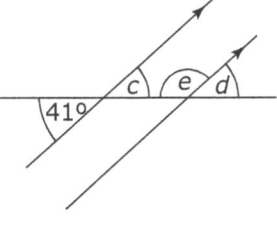

2 Calculate the unknown angles, giving reasons for your answers.

a

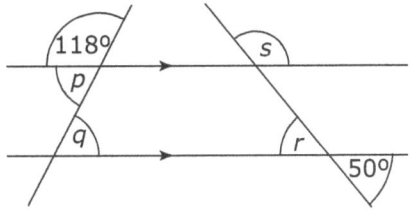

b

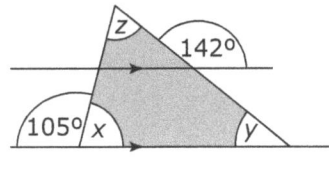

3 PQR and RST are triangles. PQ is parallel to ST. Give a reason for each answer.

a Write an angle equal to PQR.

b Write an angle equal to QPR.

c Write an angle equal to PRQ.

d What have you shown about the angles of triangles PQR and RST?

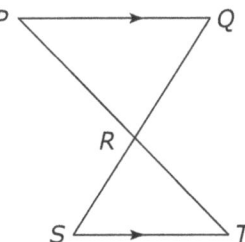

Example

Calculate the missing angles in this quadrilateral.

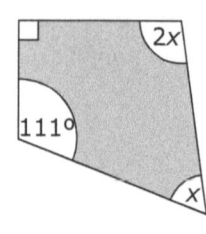

$2x + x = 360° - 90° - 111°$

$3x = 159°$

$x = 53°$

The angles are 53° and

$2 \times 53° = 106°.$

1 Calculate the unknown angles.

a
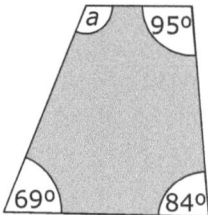

b
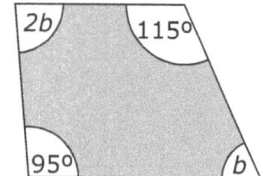

2 Calculate the angles of these quadrilaterals, giving reasons for your answers.

a
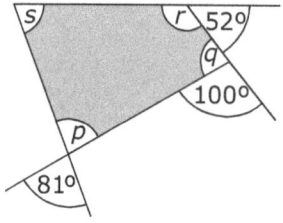

b
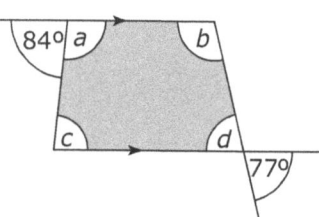

3 Calculate the unknown angles.

a
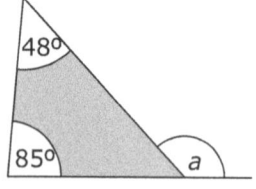

b
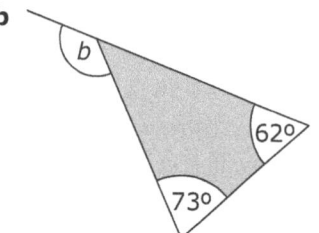

Identify the two ways in which this triangle can be classified.

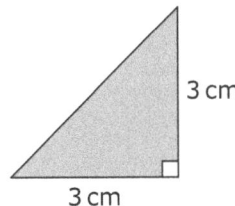

3 cm

3 cm

The triangle has two equal sides and one right angle.
This triangle is right-angled and isosceles.

1 Copy and complete this table about triangles.

Name of triangle	Number of equal sides	Number of equal angles	Number of right angles
Isosceles			0 or 1
	3		
Right-angled	0 or 2		
		0	

2 State whether each of these triangles is equilateral, isosceles, scalene or right-angled. Explain your answers.

a

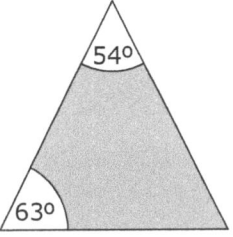

b

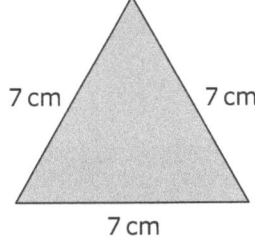

c

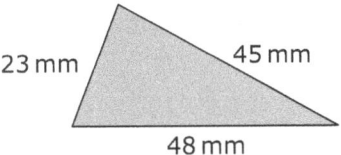

d

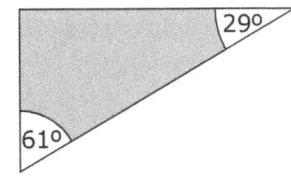

5e   Properties of quadrilaterals

Example

Sketch and name all the quadrilaterals with two different pairs of equal and opposite angles.

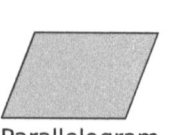

Parallelogram

Rhombus

1 Match these shapes with their descriptions and name each shape.

a Four equal sides, opposite sides are parallel, opposite angles are equal, diagonals bisect at 90° and are different lengths.

b Opposite sides are parallel and equal in length, four 90° angles, diagonals bisect each other and are equal in length.

c Opposite sides are parallel and equal in length, opposite angles are equal, diagonals bisect each other and are different lengths.

d One pair of opposite and parallel sides.

2

Square	Kite	Trapezium

Rectangle	Rhombus	Parallelogram

From the choices above, choose all the quadrilaterals that have

a four equal angles

b one pair of equal and opposite angles

c both diagonals bisected by each other

d diagonals that bisect at 90°.

MyMaths.co.uk

Q 1102   SEARCH

Example

Write the usual names of

a a regular triangle

b a regular quadrilateral.

- -

a A triangle with equal sides and equal angles is an equilateral triangle.

b A quadrilateral with equal sides and equal angles is a square.

1 **a** Write the names of these regular polygons.

i 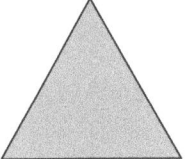 **ii**  **iii**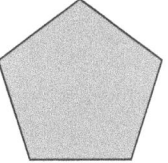

b Write the number of lines of symmetry of each shape.

c Write the order of rotation symmetry of each shape.

2 This regular pentagon has been divided into three triangles by drawing the diagonals from one vertex.

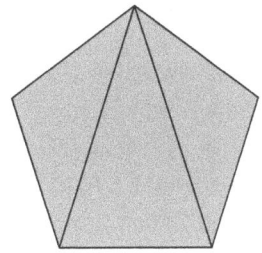

Copy and complete.

a The sum of the interior angles of a pentagon is 3 × 180° = ☐°.

b One interior angle of a regular pentagon is ☐° ÷ 5 = ☐°.

c Use this method to work out the interior angle of a regular hexagon.

Example

Three vertices of an isosceles trapezium are A (−1,2), B (2,2) and C (3, 0). Plot a fourth point, D, to complete the quadrilateral.

a Write the coordinates of point D.

b Find two pairs of equal angles.

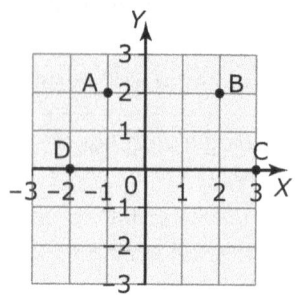

a (−2,0)

b Angle ADC = angle BCD and

angle DAB = angle CBA.

1 Draw a set of axes with values from −3 to 3.
   a Plot and join the points (3,0), (1,−3), (−1,0) and (1,3).
   b Name the quadrilateral that you have drawn.
   c Write the coordinates of the intersection of the diagonals.

2 Draw a set of axes with values from −5 to 5.
   a Plot the points (2,0), (−1,−2) and (−4,0) and do not join them.
   b If you joined the points, what type of triangle would you have drawn?
   c Plot a fourth point to create a rhombus. What are the coordinates of this point?

3 Draw a set of axes with values from −3 to 3.
   a Plot the points (−1,2) and (3,2) and join them with a straight line. Write the coordinates of the midpoint of this line segment.
   b Plot the points (−2,−3) and (−2,2) and join them with a straight line. Write the coordinates of the midpoint of this line segment.
   c Write a method to find the coordinates of the midpoint of a line segment.
   Use your method to find the coordinates of the midpoint of the line segment joining the points (−5,7) and (1,−5).

MyMaths.co.uk

Q 1093   SEARCH

Write two lines that intersect at the point $(-1, 4)$.

One line that passes through $(-1, 4)$ is the straight line made up of points whose x coordinates are all -1. For example, $(-1, 0)$, $(-1, 1)$, $(-1, 2)$, $(-1, 3)$, $(-1, 4)$. This is the line $x = -1$.

Another line that passes through $(-1, 4)$ is the straight line made up of points whose y coordinates are all 4. For example, $(-1, 4)$, $(0, 4)$, $(1, 4)$, $(2, 4)$, $(3, 4)$. This is the line $y = 4$.

1 Draw a set of axes with values from -5 to 5 and plot these graphs.

   a $x = 3$ **b** $y = 2$ **c** $x = -4$

   d $y = -1$ **e** $x = 0$

2 **a** Write the coordinates of the point where the lines $x = -1$ and $y = 2$ intersect.

   b Write what you notice about the value of the x coordinate and the value of the y coordinate when you compare these coordinates with the equations of the lines.

   c The point of intersection of these two lines is the top left-hand corner of a square of area 9 units². Write the equations of the two lines needed to complete this square.

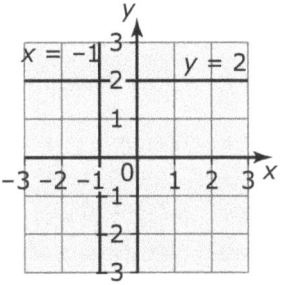

3 Without drawing the graphs, write where these lines intersect.

   a $x = 3$ and $y = 1$ **b** $x = 2$ and $y = -3$ **c** $x = -5$ and $y = -4$

4 Write two lines that intersect at these points.

   a $(2, 6)$ **b** $(-3, -8)$ **c** $(-1, 7)$

Example

Generate three coordinates that lie on the graph of $y = 3x + 1$.

Substitute $x = 1$ $y = (3 \times 1) + 1 = 4$

Substitute $x = 2$ $y = (3 \times 2) + 1 = 7$

Substitute $x = 3$ $y = (3 \times 3) + 1 = 10$

The coordinates are $(1, 4)$, $(2, 7)$ and $(3, 10)$.

1 a Copy and complete these tables to generate three coordinates for each of these line graphs.

i $y = 3x + 2$

x	1	2	3
y			

ii $x + y = 4$

x	1	2	3
y			

b Draw each graph on a separate set of axes.

2 a For each linear function, copy and complete the coordinates.

   i $y = 2x + 3$ $(-2, \square)$ $(\square, 3)$ $(2, \square)$
   ii $y = -x$ $(-3, \square)$ $(\square, -1)$ $(2, \square)$

b Plot the graphs of $y = 2x + 3$ and $y = -x$ on the same axes.

c Write the coordinates of the point where these two lines intersect.

3 Matt and Damon are builders. Matt charges a fee of £80 plus £30 for each hour that he works. Damon charges a fee of £100 plus £20 for each hour that he works.

a Draw a set of axes with x values from 0 to 8 (representing the number of hours worked) and y values from 0 to 320 in steps of 20 (representing the total cost of the job).

b Plot a line graph for each builder.

c Which builder would you use for the job if the time taken was

   i   1 hour   **ii**   2 hours   **iii**   6 hours?

MyMaths.co.uk

1395, 1396   SEARCH

Generate two points on the graph of $x - 3y = 6$.

We can let $x = 0$ and $y = 0$, in turn, to generate two points.

When $y = 0$, $x = 6$ When $x = 0$, $-3y = 6$

                                   Dividing by -3, $y = -2$

So the two points $(6, 0)$ and $(0, -2)$ are on the graph.

1 Match these graphs with their equations.

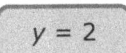 $y = 2$ $y = -3$ $y = x + 1$

$x = 0$ $x = 2$ $y = x + 3$

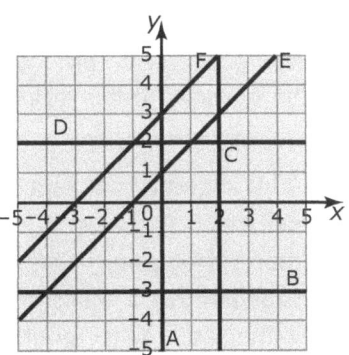

2 **a** Copy and complete these tables to generate three coordinate pairs
for each line graph.

i $y = 2x + 2$

x	1	2	3
y			

ii $y = x + 3$

x	1	2	3
y			

b Draw the graphs above on the same set of axes.

c Write the point of intersection of the two graphs.

3 **a** For each equation, copy and
complete this table of values.

x	0	
y		0

i $x + y = 5$ **ii** $x + 2y = 6$ **iii** $2x + 3y = 6$

b Draw a set of axes with values from -8 to 8 and plot these
three graphs.

⊕ MyMaths.co.uk

Example

Describe Sarah's car journey shown by this distance–time graph.

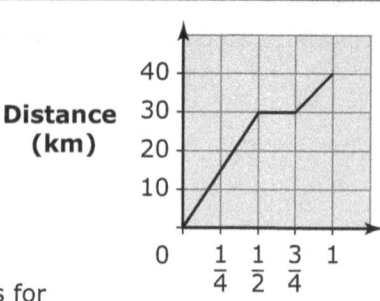

Distance (km)

- - - - - - - - - - - - - - - - - - - -

Sarah sets off and drives for 30 km during the first half an hour. She is therefore driving at an average speed of 60 km per hour.

She stops for 15 minutes and then drives for another 15 minutes, covering a distance of 10 km. She is now driving at an average speed of 40 km per hour. Her journey finishes 40 km away from her starting point.

1 Aura shows this distance–time graph to her friend Moggie and explains that it shows her journey to her Grandmother's house. Moggie immediately tells Aura that this is impossible. How does Moggie know?

Distance (km)

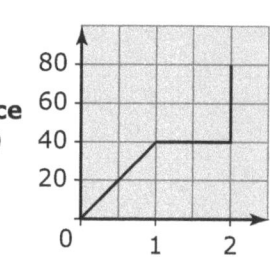

Time (hours)

2 Plot a graph of 'Distance from home' against 'Time' for this journey.

Jamie leaves his house at 7 a.m. for work. He walks 6 km in one hour and then stops at a café. He takes half an hour to eat his breakfast and then walks the last 2 km to work, arriving at 9 a.m. He spends 15 minutes trying to get in to his building before realising it is Sunday. At this point Jamie borrows an onlooker's bicycle and cycles home in half an hour.

a What time did Jamie arrive back home?

b At what speed did he cycle on his return journey?

3 £10 is currently worth 18 New Zealand dollars. Construct a conversion graph to convert British pounds to New Zealand dollars and vice versa. Go up to £50. Use your graph to convert

a £25 to New Zealand dollars

b $27 to British pounds.

MyMaths.co.uk

Q 1184   SEARCH

This table gives information about the average class size in England at Key Stage 2 during the period 1998 to 2005.

1998	1999	2000	2001	2002	2003	2004	2005
28.3	28.4	28.3	27.9	27.4	27.3	27.2	27.3

Draw a time series graph and comment on any trends shown.

The average class size has decreased on the whole, although not dramatically.

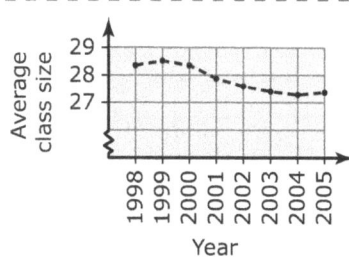

1 A waiter kept a record of his tips over a 12-month period, in pounds per month.

Jan	Feb	Mar	Apr	May	Jun	Jul	Aug	Sep	Oct	Nov	Dec
40	40	48	70	75	92	100	120	85	56	45	104

a Draw a time series graph to represent this set of data.

b Comment on the trends shown.

2 The table shows the temperature in °C taken each hour on 8 July 2007 at Wick and Camborne.

Time	10.00	11.00	12.00	13.00	14.00	15.00	16.00	17.00
Wick	12	13	13	14	14	14	15	15
Camborne	17	17	15	16	16	15	17	16

a Plot line graphs for this data on the same set of axes, remembering to join the points with broken lines.

b Describe the behaviour of the temperature in Wick and Camborne on 8 July 2007.

c Explain why your graph cannot give the temperature in Wick at 14.30.

Example

Work out   **a** 1.7×0.452 **b** $13 \div 17$
Give your answers to 2 decimal places.

a $1.7 \times 0.452 = 0.7684$ The 8 tells you to round up.

So $1.7 \times 0.452 = 0.77$ (2dp)

b $13 \div 17 = 0.764\ 705\ 882\ \ldots$ The 4 tells you to round down.

So $13 \div 17 = 0.76$ (2dp)

---

1   Round each number to the nearest 10.

a	22	**b**	47	**c**	35	**d**	99
e	124	**f**	386	**g**	505	**h**	1523

2   Round each number to the nearest
   i   100   **ii**   1000.

a	672	**b**	949	**c**	1264	**d**	5889
e	3614	**f**	7250	**g**	28410	**h**	65525

3   Round each number to the nearest tenth.

a	2.83	**b**	4.17	**c**	3.64	**d**	0.79
e	16.22	**f**	6.45	**g**	21.045	**h**	0.95

4   Round each number to the number of decimal places
   given in the question.

a	1.35 (1dp)	**b**	0.268 (2dp)
c	17.6722 (2dp)	**d**	0.559 (1dp)
e	24.081 (1dp)	**f**	54.6575 (3dp)
g	37.595 (2dp)	**h**	8.4996 (3dp)

5   Convert each fraction into a decimal using a calculator.
   Give your answer to the nearest hundredth or use
   recurring decimal notation where appropriate.

   a $\dfrac{4}{9}$ **b** $\dfrac{7}{16}$ **c** $\dfrac{23}{32}$ **d** $\dfrac{8}{11}$

Example

a   Insert brackets to make $9^2 - 21 \div 3 = 20$ true.

b   What is the correct answer if no brackets are inserted?

- -

a $(9^2 - 21) \div 3 = 20$

b $9^2 - 21 \div 3 = 81 - 21 \div 3 = 81 - 7 = 74$

1   Calculate

    a $2 \times 3 + 6$ **b** $7 + 8 \div 2$

    c $2 + 3 \times 5$ **d** $16 - 3 \times 4$

    e $2 \times 5 + 6 \times 4$ **f** $24 - 9 \div 3$

    g $4 + 18 \div 2$ **h** $14 \div 2 - 18 \div 3$

2   Calculate

    a $4 \times (2 + 8)$ **b** $(16 - 8) \div 4$

    c $(30 - 12) \times 2$ **d** $3 \times (9 - 4) \times 4$

    e 5×2^3 **f** $4^2 + 21 \div 3$

    g $\dfrac{6^2 + 4}{8}$ **h** $\dfrac{8 + (3^2 - 4)}{10}$

3   Copy these calculations and insert brackets, if necessary, to make the answers correct.

    a $3 \times 9 + 2 = 33$ **b** $15 - 2 \times 3 = 9$

    c $3 \times 5 + 2 \times 4 = 84$ **d** $8 + 24 \div 4 = 8$

    e $8 \times 8 - 4 \times 3 = 52$ **f** $100 - 36 \div 4 = 16$

    g $2^2 + 20 \div 5 = 8$ **h** $30 \div 6 - 2^2 = 15$

4   Use a calculator to work out these calculations. Where appropriate, give your answer to 2 decimal places.

    a $(11 + 2.8) \div 3.1$ **b** $(3 - 2.8)^2 \times 1.56$

    c $\dfrac{14}{9 \times 8}$ **d** $26 - 4.3^2 \div 3$

Example

George sponsors Cristina 15p for every word that she spells correctly in a sponsored spelling test. Cristina spells 49 words correctly. How much does George owe her?

You need to calculate 15p × 49.
Using compensation you calculate 15 × 50 − 15 = 750 − 15 = 735.
George owes Cristina £7.35.

1 Calculate these using factors.

a 8 × 50	**b** 16 × 40	**c** 14 × 300	**d** 1.5 × 60
e 3.2 × 30	**f** 4.4 × 500	**g** 2.25 × 40	**h** 1.75 × 800

2 Calculate these using the method of partitioning.

a 11 × 3.8	**b** 14 × 5.1	**c** 15 × 2.4	**d** 23 × 1.5
e 459 ÷ 9	**f** 328 ÷ 4	**g** 504 ÷ 6	**h** 504 ÷ 7

3 Calculate these using the method of compensation.

a 22 × 9	**b** 35 × 19	**c** 28 × 21	**d** 17 × 31
e 2.7 × 11	**f** 4.3 × 19	**g** 29 × 0.8	**h** 49 × 6.2

4 Use an appropriate mental method to calculate these.

a Meredith is setting out the chairs for assembly. She has been asked to set out 21 rows of 24 chairs. How many chairs will she have to set out?

b Derek spends £3.90 on his favourite monthly magazine. Calculate how much he spends on this magazine in one year.

c Izzie records *Dora the Explorer* on to her hard drive recorder every day in April. Each episode uses up 0.9 GB of the space. If her hard drive recorder has 160 GB of space, how much free space will she have at the end of the month?

MyMaths.co.uk

1010, 1013 SEARCH

Work out 4.2×3.5.

Change the calculation to an equivalent whole number calculation. $4.2 \times 3.5 = 42 \times 35 \div 100$

```
    42
  × 35
   210   ⟵  5 × 42
  1260   ⟵  30 × 42
  1470
```

So $4.2 \times 3.5 = 1470 \div 100 = 14.7$

1 Calculate these using long multiplication. Remember to estimate the answer first.

   a 12×27 **b** 22×39

   c 38×25 **d** 53×47

   e 123×15 **f** 218×35

   g 503×61 **h** 727×82

2 Calculate these. Remember to estimate the answer first.

   a 6×9.4 **b** 15×7.3

   c 26.2×8 **d** 3.43×46

   e 2.6×5.8 **f** 6.4×3.2

   g 4.9×2.5 **h** 8.02×7.5

3 **a** Charlie washes dishes at a hotel. He places the dishes in stacks of 24. At the end of his shift he has 18 stacks of dishes. How many dishes has he washed?

   b A bed and breakfast charges £45 per person per night. Rebecca decides to stay for two weeks. How much will her stay cost?

   c Gino charges £7.25 for one of his speciality cocktails. Each member of a party of 32 people drinks two of these cocktails. How much should Gino charge the hosts of this party?

Example

Work out 73.6 ÷ 8 using long division.

```
      9.2
  8)73.6
     72
     1.6
     1 6
        0
```

1 Use long division to calculate these. All the answers are whole numbers.
  a 216 ÷ 4   **b** 272 ÷ 8   **c** 322 ÷ 7
  d 231 ÷ 11   **e** 555 ÷ 15

2 Calculate these using long division. Give each answer with a remainder.
  a 249 ÷ 8   **b** 209 ÷ 9   **c** 596 ÷ 14
  d 2864 ÷ 23   **e** 4206 ÷ 41

3 Calculate these. Give each answer as a decimal.
  a 9.36 ÷ 3   **b** 19.28 ÷ 4   **c** 59.28 ÷ 8   **d** 37.05 ÷ 5
  e 3.672 ÷ 6   **f** 19.26 ÷ 9   **g** 166.95 ÷ 7   **h** 289.52 ÷ 8

4 Calculate these. You will need to add extra zeros to help
  you complete the division. Give each answer as a decimal.
  a 25.8 ÷ 5   **b** 52 ÷ 8   **c** 6.75 ÷ 6
  d 9.3 ÷ 4   **e** 41.4 ÷ 8

5 a Philip is allocating tutors for a training session at his
      company. Each tutor should have at most 15 employees
      in his group. 252 employees work for Philip's company.
      How many tutors does Philip need?

  b Gemma shares a lottery win of £6093.45 between herself and
      14 members of her family. How much does each person receive?

  c Lily and her seven friends share the cost of a restaurant meal
      equally between themselves. If the total bill is £129.20,
      how much do they each pay?

MyMaths.co.uk

1041, 1917   SEARCH

Example

A school bus can seat 48 students. 222 students use these buses to travel to the theatre. How many buses are needed? Are all the buses full and if not, how many students are on the bus that is not full?

$222 \div 48 = 4.625$

5 buses are required but one is not full. On this bus there are $0.625 \times 48 = 30$ students.

---

1   Calculate these. Where appropriate, give your answer to 2dp.

  a $1.3^2 - 0.75$
  b $2.9 \times \sqrt{29.16}$
  c $(8.94 - 3.04)^2$
  d $18.9 - (3.4 + 0.68)^2$
  e $\dfrac{4.1 + 8.75}{\sqrt{16}}$
  f $\dfrac{(3.05 - 1.63)^2}{5}$
  g $\dfrac{8.23 + 4.6}{\sqrt{1.5 + 0.95}}$
  h $\left(\dfrac{4.8}{3.6}\right)^2$

2   Use a calculator to work these out. For each answer turn your calculator display upside down and write the word that is shown on the display. Write the sentence that all six words make.

  a $18^2 + 7 \times 3$
  b $135 + (205 + 35)^2$
  c $25^2 - \sqrt{49}$
  d $22^2 \times 10 - 22 \times 10 - 6$
  e $\sqrt{0.09} + \dfrac{0.3667}{5}$
  f $145 + \left(\dfrac{575}{2.5}\right)^2$

3   Use a calculator to solve these problems. Give your answers in a form appropriate to the question.

  a   Chocolate eggs are packed in boxes of 48. How many boxes are needed to pack 1400 eggs and how many eggs are left over?
  b   A box of 48 eggs costs £18.00. How much does each egg cost?
  c   Five school friends share a box of 48 chocolate eggs. How many eggs do they each receive and how many eggs are left?
  d   Neil eats three of his eggs in 4 minutes and 15 seconds. How long does he take to eat each egg?

Example

Aisling recorded the number of people in each of 15 teachers' cars arriving at school one Monday morning.

1 1 3 2 1 2 1 2 1 1 1 3 4 2 2

Find   **a**   the mode   **b**   the median   **c**   the range of this set of data.

- -

a   The mode is the most common result = 1

b   Order the data to give 1, 1, 1, 1, 1, 1, 1, 2, 2, 2, 2, 2, 3, 3, 4 and choose the middle result. The median is 2.

c   The range is 4 − 1 = 3

---

1   Copy and complete the table by placing each statement under the correct heading according to the type of data that it would generate.

Discrete	Continuous	Non-numerical

   a   The number of people waiting at a bus stop

   b   The colour of students' school bags

   c   The time taken to travel to school

   d   The number of heads when three coins are tossed

   e   The weight of apples picked from a tree

   f   The shoe sizes of primary school children.

2   Find   **i**   the mode   **ii**   the median   **iii**   the range of these sets of data.

   a   5, 8, 0, 8, 2   **b**   11, 15, 13, 10, 12, 19, 10, 12, 18

   c   −7, −5, 4, −1, 0, −7, −3   **d**   103, 108, 104, 104, 106, 109

3   These are some temperatures recorded at the Amundsen–Scott South Pole Station in Antarctica.

2 p.m. June 21	8 p.m. June 21	2 a.m. June 22	8 a.m. June 22	2 p.m. June 22
−50 °C	−51 °C	−47 °C	−46 °C	−45 °C

Find   **a**   the median   **b**   the range of this set of data.

8b The mean

The total cost of five different types of washing-up liquid is £4.65.
Calculate the mean cost.

Mean $= \frac{£4.65}{5} = £0.93$

1 The total rainfall (in mm) recorded each March from 1998
to 2007 is listed below. Calculate the mean rainfall for March
during this period.

1998	1999	2000	2001	2002	2003	2004	2005	2006	2007
96.5	70.9	38.0	91.3	49.3	38.9	53.9	52.0	93.5	64.5

2 Niall recorded the time taken for a group of students
to complete a sudoku puzzle. The times in minutes are

Boys   34,   23,   19,   28,   32
Girls   27,   36,   28,   25,   30

   a Calculate the mean time taken for the boys.
   b Calculate the mean time taken for the girls.
   c Calculate the mean time taken for the whole group of students.

3 Caiomhe recorded the time taken by five of her friends to travel to
school one Friday morning.
The times in minutes are 11, 12, 9, 10, 58.
   a Calculate the mean time taken to travel to school.
   b Is this value a fair indicator of the time taken to travel to school?
      Explain your answer.

4 Orla scored these results in her end
of year examinations.
   a Calculate her mean score in maths,
      English and science.
   b When Orla receives her Spanish
      result, her mean score increases to
      75%. What is Orla's Spanish result?

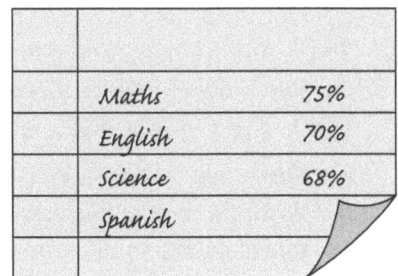

Maths	75%
English	70%
Science	68%
Spanish	

Example

The frequency table shows the numbers of letters in the first sixteen words of *Northern Lights* by Philip Pullman.

Number of letters	2	3	4	5	6	7	8	9
Frequency	2	4	5	2	1	1	0	1

Calculate the mean number of letters.

The total number of letters is

$(2 \times 2) + (3 \times 4) + (4 \times 5) + (5 \times 2) + (6 \times 1) + (7 \times 1) + (9 \times 1)$

$= 4 + 12 + 20 + 10 + 6 + 7 + 9$

$= 68$

The mean number of letters is $\frac{68}{16} = 4.25$

1 A group of children conduct a survey on the number of doors on the cars that pass their school gate in a 20-minute period. These are their results.

2 4 3 3 2 5 2 4 3 4 5 4 5 3 5 5 4 2 3 5 3

   a Copy and complete this frequency table.

Number of doors	2	3	4	5
Frequency				

   b Write the two modes.
   c Calculate the range.
   d Calculate the median number of doors.
   e Calculate the mean number of doors.

2 These are the numbers of letters in the first twenty words of the children's novel *Anne of Green Gables* by L. M. Montgomery.

3 6 5 5 4 5 3 7 4 4 6 4 4 1 6 6 7 4 6 3

   a Show this information in a frequency table.
   b Write the modal number of letters.
   c Calculate the median number of letters.
   d Calculate the mean number of letters.

MyMaths.co.uk

🔍 1202, 1254 SEARCH

The table shows where people consult their doctor. The numbers are percentages.

	1975	2005
Surgery	78	87
Telephone	3	10
At home	19	3

a   Draw a comparative bar chart to represent this information.

b   Describe the change from 1975 to 2005.

a

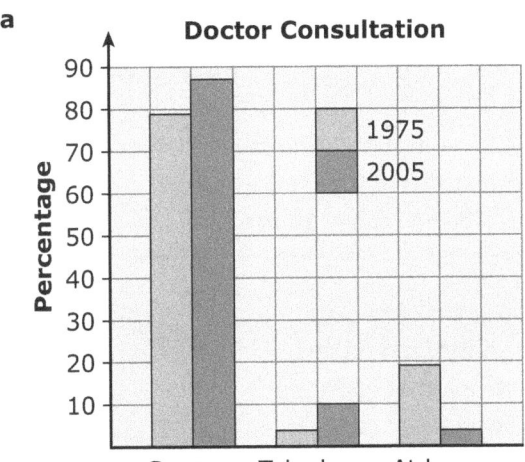

b   In 1975, it was much more commonplace for doctors to make home visits. In 2005, telephone calls seem to have replaced some of these home visits and more people make the journey to the surgery.

1   The table gives the average cost, in pence, of some basic household items in 1996 and 2006.

	1996	2006
250g Cheddar cheese	115	142
12 size 2 eggs	158	181
800g white bread	55	81
1 pint milk	36	35
1kg sugar	76	74

a   Draw a comparative bar chart to represent this information.

b   Compare the cost of these basic household items over this ten-year period.

Example

The pie chart shows the dessert choices of some friends at a party.
Eight friends chose the chocolate mousse.
Work out the number of friends that chose the other two options.

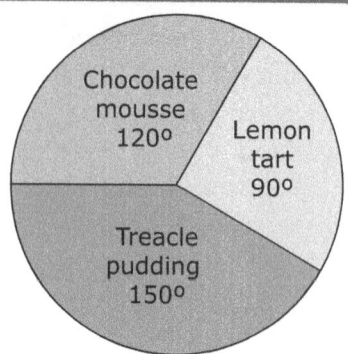

120° represents 8 friends so 120° ÷ 8 = 15° represents 1 friend.
Lemon tart was chosen by 90° ÷ 15° = 6 friends.
Treacle pudding was chosen by 150° ÷ 15° = 10 friends.

1 This pie chart shows the proportion of each ingredient required to make shortbread fingers.
   a Write the fraction of the recipe that is butter.
   b Write the fraction of the recipe that is sugar.
   c Doris wants to make 450 g of this mixture. Calculate the amount of each ingredient she needs.

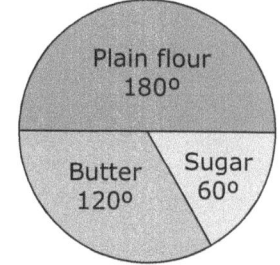

2 The GCSE maths results for St Clare's School are given in this pie chart.
   a Work out the angle representing 'Grade C'.
   b 24 students achieved a grade B. Work out the number of students taking GCSE maths in St Clare's School.
   c Calculate the number of students who achieved an A*.

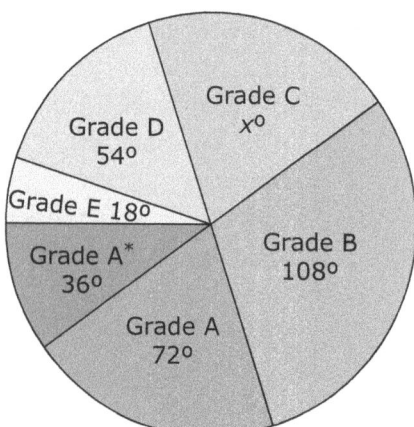

Mrs Hinds wants to know whether the students in her school believe they are getting the right amount of homework each night. Design a data collection sheet for Mrs Hinds to use.

Amount of homework				
Name	**Too little**	**Just right**	**Too much**	**Not sure**

1 Mark wants to plant a hedge. He decides to investigate different varieties and base his decision on how tall the hedge will grow and whether or not it is evergreen. Design a data record sheet for Mark to use.

2 Nicky is buying a new car. She wants to draw up a data collection sheet on which she can record the features that are important to her. These are the size of the engine, the fuel type, the transmission (manual or automatic) and whether or not the car has a full service history.
   a Design a data record sheet on which Nicky can record the cars she has seen.
   b Suggest two other features that she may like to add to her data record sheet.

3 Lauren wants to know how different students travel to school and whether age makes a difference to their choice of transport. Design a data collection sheet for her to use.

4 Callum is planning to organise a trip to a leisure centre for all his friends. He decides to find out the available facilities at all his local centres in order to decide the best one to go to. Create a data record sheet for Callum to use.

8g   Designing a questionnaire

Write what is wrong with this question and suggest improvements.

> *Which is your favourite flavour of crisps?*
> ☐ *Cheese and onion*   ☐ *Prawn cocktail*

This question will leave some people feeling that they were unable to give their opinion. More categories should be added. For example

> ☐ *Salt and vinegar*   ☐ *Ready salted*
> ☐ *Beef*   ☐ *Other*

1   For each survey, write why the place chosen for conducting the survey may bias the results.

   a   To find out the popularity of school dinners by asking students as they leave the school canteen.

   b   To find out the preferred type of music by asking members of the public as they leave a Kylie Minogue concert.

   c   To find out the preferred brand of toilet paper by asking all the people leaving a superstore at 2 p.m. on a Monday.

2   Rupert writes this survey to find out opinions on films.

> *1. Don't you think that comedy is the best type of film?*
>
> *2. How often do you buy films on DVD?*
> ☐ *A lot*   ☐ *A little*
>
> *3. How old are you?*
> ☐ *10 and under*   ☐ *10–40*   ☐ *Over 50*

   a   Write what is wrong with each of these questions.

   b   Improve and rewrite each question.

   c   Add another question of your own that would improve this survey.

This is a grouped frequency table showing the rainfall, *r*, in millimetres during June from 1998 to 2007.

a What does the group 150 ≤ *r* < 180 mean?

b State the modal class.

Rainfall, *r*, in mm	Frequency
0 ≤ *r* < 30	3
30 ≤ *r* < 60	9
60 ≤ *r* < 90	3
90 ≤ *r* < 120	2
120 ≤ *r* < 150	2
150 ≤ *r* < 180	1
Total	20

a This means that the rainfall is more than or equal to 150 mm and less than 180 mm.

b The modal class is 30 ≤ *r* < 60.

1 These are the heights, *h*, in cm of a group of 30 Year 7 girls.

145.4 147.0 142.3 150.6 148.5 152.5 141.0 149.3 152.6
147.8 143.3 142.4 154.9 150.8 151.4 139.2 146.7 138.5
145.6 144.7 146.9 148.8 150.6 153.0 146.2 144.1 147.0
158.3 149.8 143.9

a Copy and complete this grouped frequency table.

Height, *h*, in cm	Tally	Frequency
135 ≤ *h* < 140		
140 ≤ *h* < 145		
145 ≤ *h* < 150		
150 ≤ *h* < 155		
155 ≤ *h* < 160		
	Total	

b State the modal class.

c This group of Year 7 girls are going on a trip to Alton Towers. They are all keen to ride 'Oblivion' but the height restriction says you must be over 1.4 m.
How many girls will be unable to ride?

Example

This table gives information about the 200 m running times, in seconds, for two 13-year-old athletes.

	Mean	Range
Tristan	24.79	2.14
Simon	25.15	1.73

Compare Tristan's and Simon's performances.

Tristan has the lowest mean time so he is, on average, the faster athlete. However, Simon has a smaller range, so he is more consistent.

1 Twins Alex and Imogen are arguing over their examination results. Each twin believes that they have achieved the most noteworthy scores. Here are their percentages.

	Maths	English	Science	Spanish	History	Art
Alex	73	84	54	65	94	50
Imogen	74	75	70	64	77	60

a Calculate
    i   the mean
    ii the range for each set of results.
b Compare the twins' results using your calculations from part **a**.

2 Julia and Ryan wait on tables in a restaurant. At the end of each night the amount that they have made in tips gives a basic indication of the quality of service that they have given. Here are their tips for one week.

Julia	£28	£16	£43	£12	£35	£30	£11
Ryan	£28	£25	£30	£29	£32	£27	£25

a Calculate the i   mean   ii   median   iii   mode
    iv   range for each person's data.
b A promotion is available for one person. Which average would you use to demonstrate the quality of service that you give if you were
    i   Julia   ii   Ryan?

⊕ MyMaths.co.uk

Q 1192   **SEARCH**

a Draw a set of axes with x values from -5 to 5 and y values from 0 to 5. Plot the points $(0,0)$, $(0,3)$, $(3,3)$ and $(5,0)$.

b Reflect this shape in the y-axis and name the resulting shape.

c List the coordinates of the four vertices of the resulting shape.

---

a

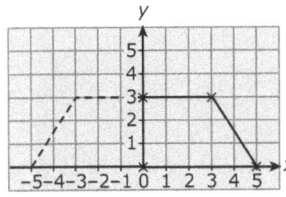

b The shape is an isosceles trapezium.

c The vertices are $(-5,0)$, $(-3,3)$, $(3,3)$ and $(5,0)$.

1 Copy each diagram and reflect each shape in the mirror line.

a   **b**   **c**

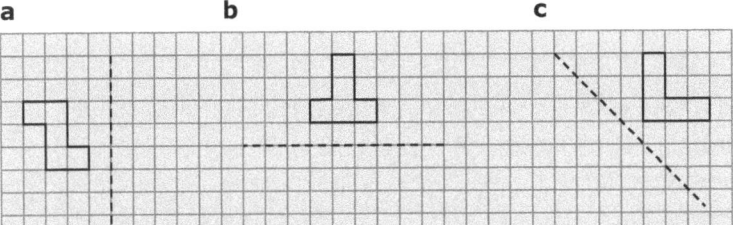

2 Copy this diagram.

a Reflect the shape in line 1.

b Reflect both the original shape and the image from part **a** in line 2.

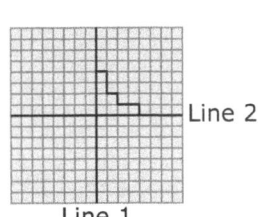

Line 2

Line 1

3 Draw a set of axes with x and y values from -5 to 5.

a Plot and join the points $(1,1)$, $(2,3)$, $(5,3)$ and $(4,1)$.

b Name the shape that you have drawn and label it P.

c Reflect this shape in the x-axis and label its image A.

d Write the coordinates of shape A.

e Reflect shape P in the y-axis and label its image B.

f Write the coordinates of shape B.

a Plot and join the coordinates $(1,0)$, $(1,3)$, $(2,3)$, $(2,1)$, $(3,1)$ and $(3,0)$.
Label this shape L_1.

b Rotate L_1 through 90° anticlockwise about $(0,-1)$ and label the image L_2.

a, b

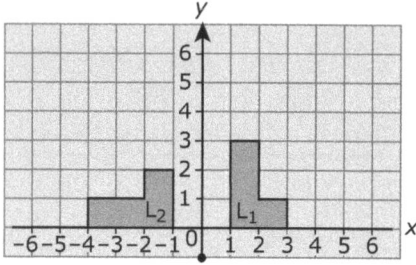

1 Copy and rotate each shape through the given angle where C is the centre of rotation.

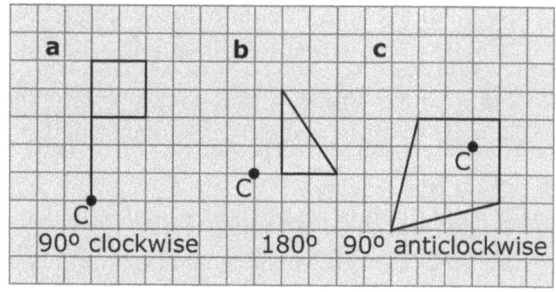

a	b	c
90° clockwise	180°	90° anticlockwise

2 Draw a set of axes with x values from -6 to 6 and y values from 0 to 6.

a Plot and join the coordinates $(2,2)$, $(4,2)$ and $(2,5)$. Label this triangle T_1.

b Rotate triangle T_1 through 90° clockwise about $(2,2)$ and label the image T_2.

c Rotate triangle T_2 through 180° about $(0,2)$ and label the image T_3.

d Describe the rotation that will take triangle T_1 directly to triangle T_3. Include the coordinates of the centre of rotation and the angle.

Copy and complete this shape so that it has rotation symmetry of order 4 about the centre, C.

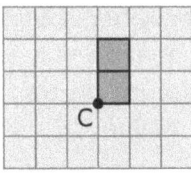

You need to rotate the shape 90° clockwise, 180° and 90° anticlockwise.

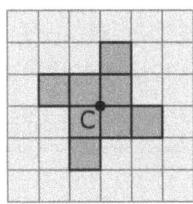

1 a Copy these road signs and draw the lines of symmetry, if possible.

b For each road sign, state the order of rotation symmetry.

2 a Copy these shapes and draw the lines of symmetry, if possible.

b For each shape, state the order of rotation symmetry.

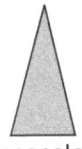

Isosceles triangle

Parallelogram

Regular pentagon

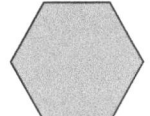

Regular hexagon

3 a Copy this diagram. Using the centre of rotation, C, complete the shape so that it has rotation symmetry of order 2.

b Copy the diagram again. Using the centre of rotation, C, complete the shape so that it has rotation symmetry of order 4.

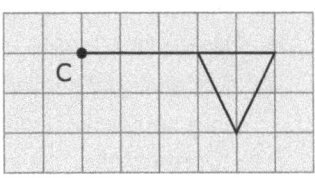

a Translate rectangle A 4 units to the right and 3 units down.

b Describe the translation that returns rectangle A to its original position.

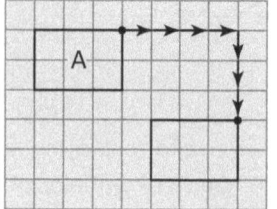

a Note how the top right-hand vertex is still the top right-hand vertex after the translation.

b 4 units to the left and 3 units up.

---

1 When a triangle is translated, point A on the triangle translates to point B on the image.

    a Describe the translation of point A to point B.

    b Copy the triangle and complete the image.

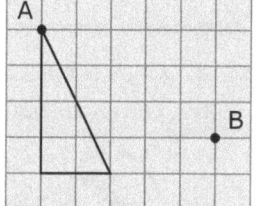

2 Describe these translations.

a	A to B	**b**	B to A
c	B to D	**d**	C to E
e	E to B	**f**	A to E
g	D to C	**h**	C to A

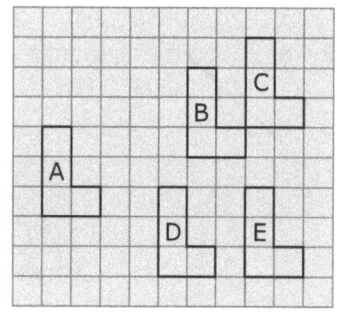

3 Draw a set of axes with values from −8 to 8.

    a Plot and join the coordinates (1, 1), (2, 4), (3, 4) and (2, 1).

    b Name this shape and label it P.

    c For each of the points in part **a**, add 2 to the x-coordinate and subtract 3 from the y-coordinate. Write the new coordinates.

    d Plot and join these new coordinates and label the shape Q.

    e Describe the translation of shape P to shape Q.

MyMaths.co.uk

9e Enlargement

Example

Triangle T_2 is an enlargement of triangle T_1. Calculate the scale factor.

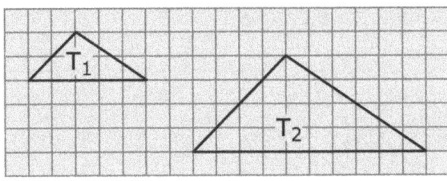

- - - - - - - - - - - - - - - -

All lengths on triangle T_2 are twice those on T_1. The scale factor is 2.

1 a Which of these shapes are enlargements of triangle T?

   b Write the scale factor of each enlargement.

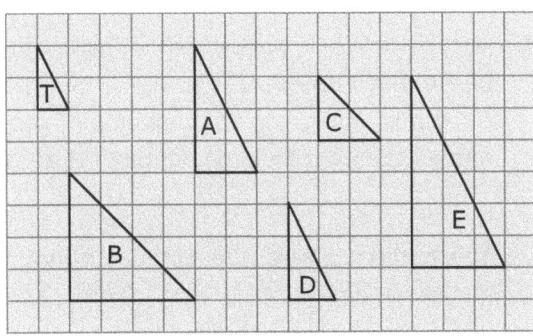

2 a Which of these shapes are enlargements of rectangle R?

   b Write the scale factor of each enlargement.

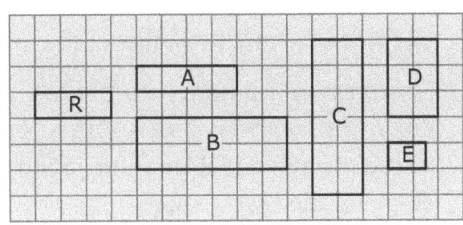

3 Copy and enlarge these shapes by the scale factor given.

   a Scale factor 2
   b Scale factor 4
   c Scale factor 2
   d Scale factor 3

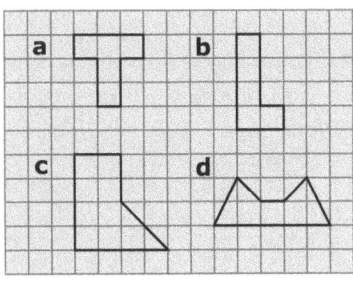

Example

Tessellate this shape six times.

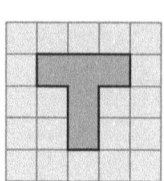

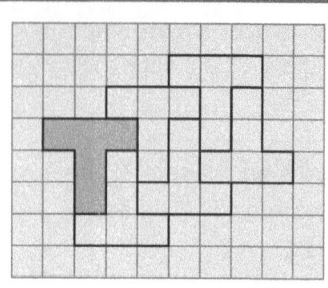

1 This pattern is a parquet wood flooring in a herringbone design.

Copy this herringbone tessellation of rectangles and add 6 more tiles to continue the tessellation.

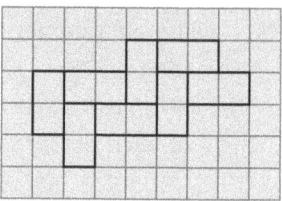

2 This pattern is a parquet wood flooring in a chevron design.

Copy this chevron tessellation of parallelograms and add 6 more tiles to continue the tessellation.

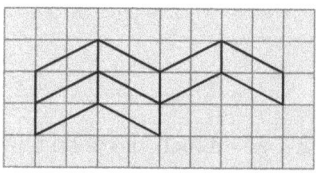

3 This pattern is a honeycomb made by bees. Copy this honeycomb tessellation of hexagons and add 6 more hexagons to continue the tessellation.

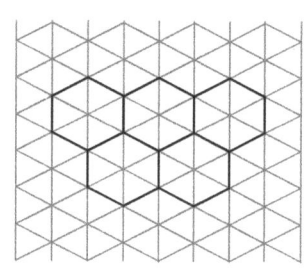

4 Copy these tessellations using pentominoes and add 6 more tiles to continue the pattern.

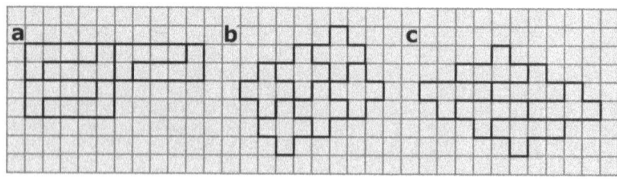

Solve these equations.

a $g + 12 = 20$ **b** $3 \times t = 12$ **c** $\frac{d}{2} = 5$

---

a $g + 12 = 20$

$g + 12 - 12 = 20 - 12$

$g = 8$

b $3 \times t = 12$

$t \times 3 = 12$

$t \times 3 \div 3 = 12 \div 3$

$t = 4$

c $\frac{d}{2} = 5$

$\frac{d}{2} \times 2 = 5 \times 2$

$d = 10$

Solve these equations: **a** $2x + 3 = 11$ **b** $\frac{(y-1)}{5} = 3$

a $2x + 3 = 11$, therefore $2x = 11 - 3 = 8$, so $x = 8 \div 2 = 4$

b $\frac{(y-1)}{5} = 3$, therefore $y - 1 = 3 \times 5 = 15$, so $y = 15 + 1 = 16$

1 Use inverse operations to find the value of each unknown.

a
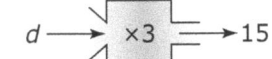
$d \longrightarrow \boxed{\times 3} \longrightarrow 15$

b
$g \longrightarrow \boxed{+3} \longrightarrow 19$

c
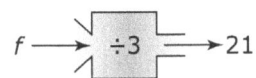
$f \longrightarrow \boxed{\div 3} \longrightarrow 21$

d
$h \longrightarrow \boxed{-3} \longrightarrow 6$

2 Solve these equations.

a $t + 5 = 15$ **b** $r - 4 = 10$ **c** $f + 12 = 20$

d $g + 21 = 45$ **e** $v + 23 = 50$ **f** $p - 3 = 33$

3 Solve these equations.

a $10 + y = 20$ **b** $16 + d = 21$ **c** $f - 14 = 2$

d $16 + u = 23$ **e** $q - 15 = 19$ **f** $k - 3 = 3$

4 Find the value of the unknown in each of these equations.

a $4 \times h = 12$ **b** $12 \times t = 48$ **c** $4 \times d = 16$

d $6c = 30$ **e** $7g = 42$ **f** $6b = 48$

5 Find the value of the unknown in each of these equations.

  a $d \div 3 = 2$ **b** $t \div 4 = 5$ **c** $q \div 5 = 25$

  d $\frac{f}{3} = 7$ **e** $\frac{h}{6} = 3$ **f** $\frac{j}{7} = 7$

For questions **1** to **4** solve the equations by any appropriate method. Find which equation, **a**, **b** or **c**, has a different solution from the other two.

6 a $x + 5 = 8$ **b** $x + 11 = 15$ **c** $x + 9 = 12$

7 a $y - 5 = 12$ **b** $y - 9 = 7$ **c** $y - 12 = 4$

8 a $5z = 65$ **b** $4z = 60$ **c** $7z = 105$

9 a $\frac{t}{3} = 16$ **b** $\frac{t}{8} = 6$ **c** $\frac{t}{7} = 7$

10 Solve these equations which have negative solutions.

  a $2x + 11 = 5$ **b** $6y + 19 = 1$ **c** $\frac{z}{3} + 5 = 2$

  d $\frac{t}{4} + 7 = 2$ **e** $\frac{(u + 10)}{2} = 4$ **f** $\frac{(v + 15)}{3} = 3$

11 a This isosceles trapezium has
     a perimeter of 26 cm.
     Find the value of x.

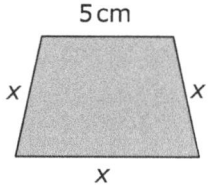

  b If the distance from London
     to Exeter is 273 km, find the
     value of x and the distance
     from Basingstoke to Exeter.

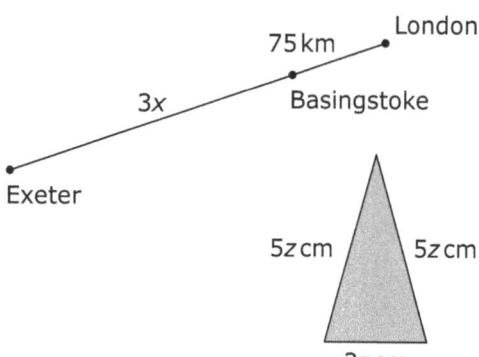

  c This isosceles triangle has
     a perimeter of 36 cm.
     Find the three side lengths.

Solve this equation.

$9(x + 2) = 5(x + 4)$

- -

$9x + 18 = 5x + 20$	Expand the brackets
$4x + 18 = 20$	Subtract $5x$ from both sides
$4x = 2$	Subtract 18 from both sides
$x = \frac{2}{4} = \frac{1}{2}$	Divide both sides by 4

1 Solve these equations.

a	$4x = 3x + 10$	**b**	$12p = 8p + 16$
c	$10k = 4k - 6$	**d**	$3t + 15 = 13t$
e	$2(n + 5) = 6n$	**f**	$6y - 18 = 3y$
g	$8(w - 6) = 2w$	**h**	$7(a + 3) = 4a$

2 Pierre is solving an equation with unknowns on both sides.
Unfortunately, he has muddled up his lines of working.
Can you help him?

$8n = 16$	$8n - 6 = 10$
$n = 2$	$12n - 6 = 4n + 10$

3 Solve these equations.

a	$5n + 7 = 3n + 11$	**b**	$4p + 8 = 10p + 2$
c	$4x + 7 = x + 16$	**d**	$3y - 4 = 2y + 3$
e	$8k + 10 = 13k - 5$	**f**	$12t - 1 = 14t - 5$

4 Solve these equations.

a	$4(x + 2) = 3x + 11$	**b**	$9p + 6 = 3(p + 8)$
c	$5(t - 2) = t + 10$	**d**	$7y - 9 = 3(y + 5)$
e	$4(k + 2) = 3(k + 5)$	**f**	$3(n + 4) = 5(n + 1)$
g	$12(b - 3) = 8(b - 2)$	**h**	$6(a + 3) = 4(a - 1)$

Example

Solve $4(x - 5) = 8 - 3x$.

$4x - 20 = 8 - 3x$	Expand the brackets
$7x - 20 = 8$	Add 3x to both sides
$7x = 28$	Add 20 to both sides
$x = 4$	Divide both sides by 7

1 Show that all these equations have a solution of $x = 3$.

   a $x + 9 = 12$ **b** $4x - 7 = 5$

   c $5(x - 1) = 10$ **d** $\frac{x}{3} + 5 = 6$

   e $7x = 2x + 15$ **f** $6(x - 2) = 2x$

   g $8x - 10 = 3x + 5$ **h** $4(x + 2) = 5(x + 1)$

2 Joe is solving an equation. Unfortunately, he has muddled up his lines of working.
Can you unscramble them for him?

$t = 2$	$7t - 3 = 11$	$5t - 3 = 11 - 2t$	$7t = 14$

3 Solve these equations.

   a $5t - 6 = 8 - 2t$ **b** $10 - 3y = 2y - 5$

   c $3(n - 9) = 13 - 5n$ **d** $7 - 4a = 5(a - 4)$

   e $2(p - 4) = 4(1 - p)$ **f** $6(k - 3) = 4(3 - k)$

4 Write an equation to describe each problem and then solve your equation to answer the problem.

   a I think of a number, double it and subtract 7. I get the same answer as when I subtract the number from 11. Find the number.

   b I think of a number, multiply it by 3 and subtract 8. I get the same answer as when I multiply the number by 2 and subtract the result from 12. Find the number.

MyMaths.co.uk

1182, 1928   SEARCH

The sum of three consecutive even numbers is 108.
Find these numbers.

If the first number is x, the next two even numbers are $x + 2$
and $x + 4$

$$x + (x + 2) + (x + 4) = 108$$
$$3x + 6 = 108$$
$$3x = 102$$
$$x = 34$$

So the numbers are 34, 36 and 38.

1 The sum of Talitha's current age and her age 5 years ago is the same as her age in 3 years' time.
   a Let Talitha's current age be x. Write an expression in terms of x for her age 5 years ago.
   b Write an expression in terms of x for her age in 3 years' time.
   c Construct an equation in x and solve it to find Talitha's age.

2 The sum of three consecutive whole numbers is 57.
   a Let the first number be x. Write an expression in terms of x for the next two consecutive numbers.
   b Construct an equation in x and solve it to find these numbers.

3 This rectangle has an area of 20 units². Write an equation to find x.

$3x - 2$

5

4 Three friends spend £500 on a shopping trip. Keira spends three times as much as Aimée who spends £100 more than Suzanne. Work out how much each friend spends.

11a Squares and square roots

Example

Write 484 as a product of its prime factors and use this to explain why 484 is a square number.

- -

$484 = 2 \times 242 = 2 \times 2 \times 121 = 2 \times 2 \times 11 \times 11 = 2^2 \times 11^2$
484 can be written as $(2 \times 11) \times (2 \times 11)$, a number multiplied by itself, which is the definition of a square number.

1 Without using a calculator, write the value of these.

    a 7^2 **b** 15^2 **c** $\sqrt{49}$ **d** $\sqrt{81}$ **e** $\sqrt{196}$

2 Use your calculator to find each of these.

    a 27^2 **b** 9.8^2 **c** $\sqrt{1156}$ **d** $\sqrt{2704}$ **e** $\sqrt{153.76}$

3 **a** Copy and continue this sequence for three more lines.

$$1 = 1$$
$$1 + 3 = 4$$
$$1 + 3 + 5 = 9$$

    b What type of numbers are on the right-hand side of each equation?

4 **a** Between which two whole numbers does $\sqrt{253}$ lie?

    b Use your calculator to find a good estimate for $\sqrt{253}$.

5 **a** Write 324 as a product of its prime factors and use this to explain why 324 is a square number.

    b Without using a calculator, find $\sqrt{324}$.

6 **a** Write 48 as a product of its prime factors.

    b Use part **a** to help you find the smallest whole number that 48 can be multiplied by to give a square number.

11b Factors and multiples

Find the LCM of 6 and 14.

List some multiples of 6:   6, 12, 18, 24, 30, 36, (42), 48, ...
List some multiples of 14:   14, 28, (42), 56, 70, ...
Choose the smallest number that appears in both lists.
The LCM of 6 and 14 is 42.

1 Write the first eight multiples of these numbers.

   a 3   **b** 5   **c** 8   **d** 12   **e** 25

2 Using your answers to question **1** to help you, find the lowest common multiple (LCM) of these pairs of numbers.

   a 3 and 5   **b** 3 and 8   **c** 8 and 12   **d** 5 and 25

3 This table shows all the products that make up the numbers 12 and 15. Copy and complete the table and find the highest common factor (HCF) of 12 and 15.

Number	Products			Factors
12	1 × 12	2 × 6	3 × 4	
15	1 × 15	3 × 5		

4 Find the HCF of these pairs of numbers.

   a 8 and 10   **b** 6 and 15   **c** 18 and 30   **d** 7 and 29

   e What can you say about the numbers in part **d**?

5 True or false? Explain your answers.

   a The smallest factor of any number is 1.

   b All numbers except 1 have an even number of factors.

   c Any multiple of 8 is also a multiple of 2 and 4.

Example

Tilly writes 120 as $p^3 \times q \times r$ where p, q and r are prime.
Find the values of p, q and r.

$120 = 2^3 \times 3 \times 5$
So $p = 2$, $q = 3$, $r = 5$

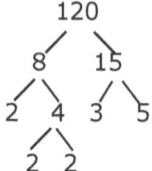

1 Copy and complete this prime factor decomposition of 24.

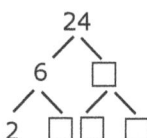

$24 = 2 \times \square \times \square \times \square$
$ = 2^{\square} \times \square$

2 Write each number as a product of its prime factors.
   a 16   b 36   c 60   d 126   e 180

3 Clara writes the prime factor decomposition of a
   number as $2^3 \times 5^2 \times 11$.
   What number is she working with?

4 The prime factor decomposition of 392 is $2^3 \times 7^2$.
   Use this fact to work out if
   a 6 is a factor of 392
   b 14 is a factor of 392.

5 108 can be written as $x^2 \times y^3$ where x and y are prime.
   Find the values of x and y.

MyMaths.co.uk

Q 1044   SEARCH

Only one of these numbers is a prime number.

151   153

a   Which is which?
b   List the factors of each number.

- -

a   If a number is divisible by 3 then the sum of its digits is divisible by 3.
For 153, 1 + 5 + 3 = 9 and so 3 is a factor of 153.
A prime number has only 1 and itself as factors so 151 must be prime.
b   The factors of 151 are 1 and 151.
The factors of 153 are 1, 3, 9, 17, 51 and 153.

1   Write all the factors of
   a   12   **b**   21   **c**   24   **d**   30   **e**   55

2   **a**   Write all the factors of
      i   4   **ii**   16   **iii**   25   **iv**   49   **v**   81
   b   The numbers in part **a** are square numbers. What do you notice about the number of factors of square numbers? Explain why.

3   A number is divisible by 9 if the sum of the digits is divisible by 9. Is 9 is a factor of these numbers?
   a   108   **b**   280   **c**   567   **d**   2549   **e**   8478

4   A number is divisible by 6 if it is an even number and the sum of its digits is divisible by 3. Is 6 a factor of these numbers?
   a   145   **b**   234   **c**   321   **d**   982   **e**   2412

5   Use divisibility tests to find out which of these numbers are prime.
   a   31   **b**   47   **c**   51   **d**   63   **e**   89

Example

Given that $132 = 2 \times 2 \times 3 \times 11$ and $252 = 2 \times 2 \times 3 \times 3 \times 7$, find

a   the HCF of 132 and 252

b   the LCM of 132 and 252.

---

a $132 = 2 \times 2 \times 3 \times 11$
$252 = 2 \times 2 \times 3 \times 3 \times 7$

Choose the prime factors that 132 and 252 have in common and multiply these together. $2 \times 2 \times 3 = 12$

b $132 = 2^2 \times 3 \times 11$
$252 = 2^2 \times 3^2 \times 7$

Choose the highest power of each prime factor in the two lists and multiply these together. $2^2 \times 3^2 \times 7 \times 11 = 2772$

---

1   From this box of numbers, write

    a   two multiples of 7   **b**   two factors of 28

    c   two square numbers   **d**   two prime numbers.

4	7	17	21	23	25

2   **a**   List the first five multiples of 3.

    b   List the first five multiples of 5.

    c   Use your lists to write the LCM of 3 and 5.

3   Find the LCM of

    a   4 and 6   **b**   8 and 10   **c**   9 and 15   **d**   2, 4 and 5

4   **a**   List the factors of 12.

    b   List the factors of 30.

    c   Use your lists to write all the common factors of 12 and 30.

    d   Write the HCF of 12 and 30.

5   Find the HCF of

    a   6 and 10   **b**   12 and 24   **c**   24 and 32   **d**   12, 15 and 45

6   Given that $126 = 2 \times 3 \times 3 \times 7$ and $396 = 2 \times 2 \times 3 \times 3 \times 11$
find   **a**   the HCF of 126 and 396   **b**   the LCM of 126 and 396.

MyMaths.co.uk

Q 1034, 1044   **SEARCH**

Use compasses to construct an angle of 45º.

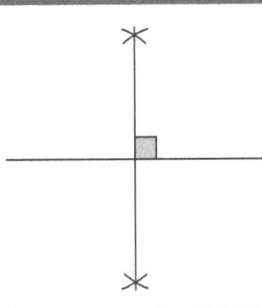

Draw an angle of 90º by constructing the perpendicular of a line.

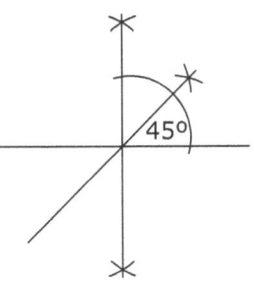

Construct the angle bisector of this angle to obtain a 45º angle.

1 a Draw a line XY, so that XY = 5.6 cm.

b Using compasses, construct the perpendicular bisector of XY and label it AB. Label the midpoint of XY as M.

c Measure the length of XM. Check that it is the same as the length MY.

d Construct the bisector of angle AMY.

2 Use compasses to construct an angle of $22\frac{1}{2}^{\circ}$.

$22\frac{1}{2}^{\circ}$ is half of 45º.

3 a Draw a line AC, so that AC = 4 cm.

b Using compasses, construct the perpendicular bisector of AC and label the points where the arcs meet B and D (with D below B).

c Join the points in alphabetical order (A to B, B to C etc.).

d Name the quadrilateral that you have constructed.

Example

From a point, X, on the ground, the angle of elevation of the top of a building is 40°. X is 12 m away from the base of the building. Find the height of the building.

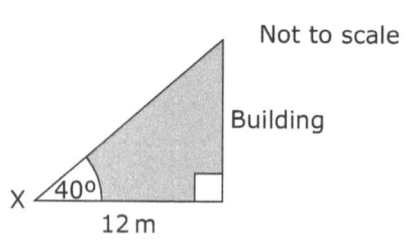

Not to scale

Building

40°

X

12 m

Make an accurate drawing where 1 cm represents 1 m.

By measuring you can see that the height of the building is 10 m.

1 Construct these triangles (ASA).

a
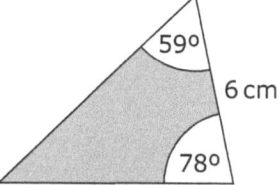
59°
6 cm
78°

b
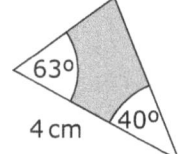
63°
4 cm
40°

2 Construct these triangles (SAS).
State the type of each triangle.

a
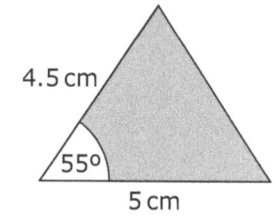
4.5 cm
55°
5 cm

b
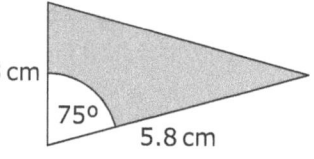
3 cm
75°
5.8 cm

3 From a point, X, on the ground, the angle of elevation of the top of a tree is 34°.
X is 7.5 m away from the base of the tree.
By making an accurate drawing where 1 cm on the diagram represents 1 m in real life, work out the height of the tree.

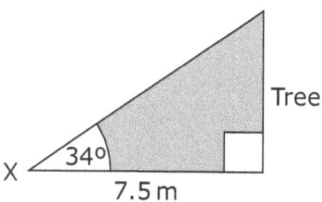

Tree

34°

X

7.5 m

MyMaths.co.uk

Q 1090   SEARCH

Construct a triangle with sides of length 3.5 cm, 3.5 cm and 5 cm.

Not to scale

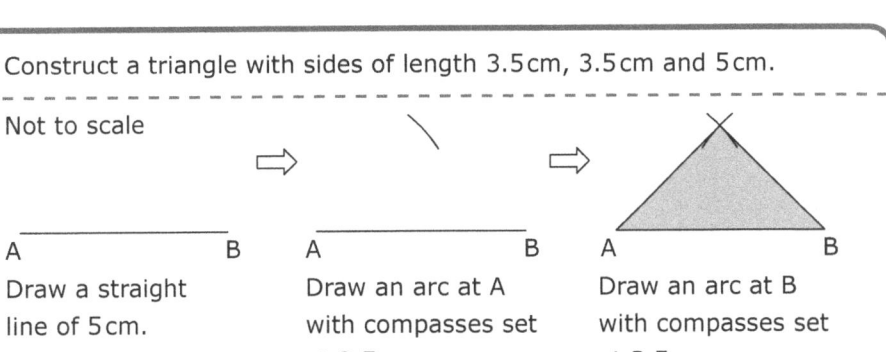

A   B   A   B   A   B

Draw a straight line of 5 cm.   Draw an arc at A with compasses set at 3.5 cm.   Draw an arc at B with compasses set at 3.5 cm.

1   **a**   Using compasses and a protractor, construct these triangles.

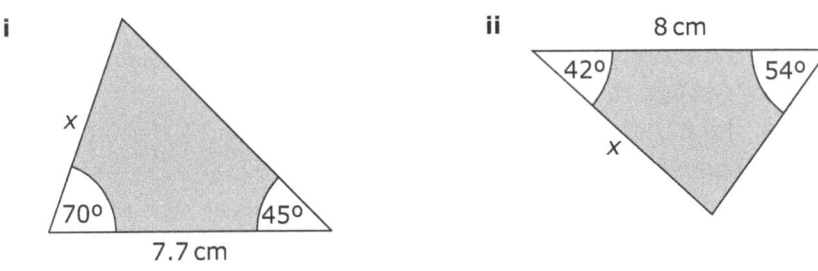

    i   **ii**   8 cm

x 42°   54°

                                        x

    70°   45°

        7.7 cm

  b   Use your ruler to measure the sides marked x.

2   **a**   Using only compasses, construct triangles with sides of length
      i   3 cm, 4 cm and 5 cm
      ii   2 cm, 5 cm and 5 cm
      iii   4 cm, 4 cm and 4 cm.
  b   Write the mathematical name of each triangle in part **a**.

3

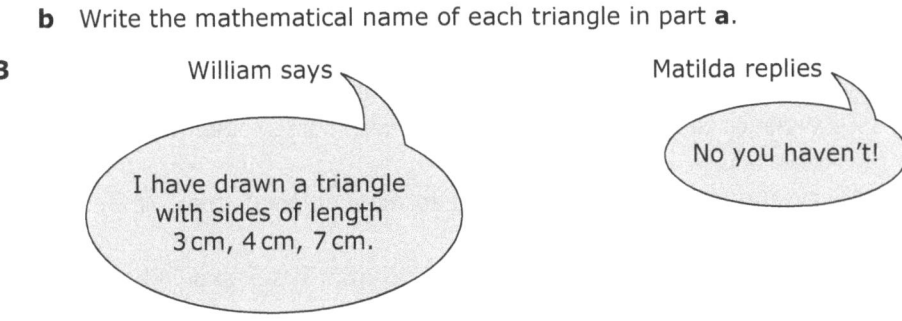

William says   Matilda replies

                                      No you haven't!

I have drawn a triangle with sides of length 3 cm, 4 cm, 7 cm.

Matilda has not looked at William's book. How can she be so sure?

Example

Draw the locus of all points that are 1 cm away from a straight line AB of length 5 cm. Describe the locus in words.

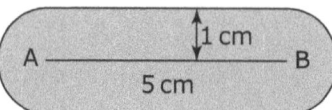

Not actual size

The locus is two lines parallel to and the same length as AB, joined at either end with semicircles whose centres are A and B.

1 Draw a sketch and describe in words the locus of

    a the tip of the minute hand on a clock as the time changes from 2 o'clock to 3 o'clock

    b the tip of the hour hand on a clock as the time changes from 2 o'clock to 3 o'clock

    c the top right-hand corner of a page of this book if you turn that page over

    d the centre of a car wheel as the car travels along a road

    e a point on the tyre of a car as the car travels along a road.

    Hint: For part **e**, mark a point on a coin and roll it along your ruler.

2 For these questions, the locus of the point will involve shading a region. Draw a sketch and describe in words the locus of

    a a garden sprinkler with a reach of 4 m

    b a dog tethered to a fence by a lead of length 3 m.

3 a Draw two points, X and Y, 6 cm apart. Draw the locus of all points equidistant from X and Y.

    Hint: Pick a few points where the distance to X is the same as the distance to Y and mark with dots. These dots should help you draw this locus.

    b On the same diagram, draw the locus of all points that are 2 cm away from Y.

    Hint: 'Plot' a few points that are 2 cm from Y. You should begin to see the shape of your locus form!

MyMaths.co.uk

Q 1147   SEARCH

12e Scale drawings

A scale model of an Aston Martin Vanquish S car has length
14.6 cm and width 6 cm. The scale used to build the model is
1 : 32 which means that for every 1 cm on the model, the real Aston
Martin uses 32 cm. Calculate, in mm, the real dimensions of an Aston
Martin Vanquish S.

14.6 cm × 32 = 467.2 cm   6 × 32 = 192 cm
To convert cm to mm × 10.
The dimensions of the real car are 4672 mm by 1920 mm.

1 This is a scale drawing of
   Carlos's bedroom. The scale is
   1 cm represents 50 cm.
   a Measure the length and
      width of Carlos's bedroom
      on the plan.
   b Write the actual length and
      width of Carlos's bedroom.
   c What is the width of Carlos's
      desk?
   d Carlos wants to hang fairy
      lights across one of the diagonals
      of his room. What is the minimum
      length of fairy lights that he requires?

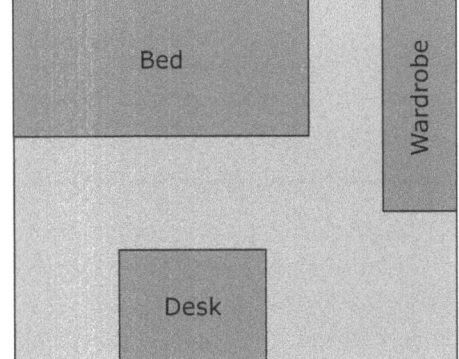

2 A scale model of a Jaguar XK8 car is 20.0 cm long.
   The scale used to build the model is 1 : 24 which means that for
   every 1 cm on the model, the real Jaguar uses 24 cm.
   a Work out the actual length of a Jaguar XK8.
   b The actual width of a Jaguar XK8 is 2071 mm. Work out how
      wide the model should be. Give your answer to 1 decimal place.

Example

Sketch two different ways in which a triangular prism can be cut to leave two identical prisms. Describe the resulting prisms.

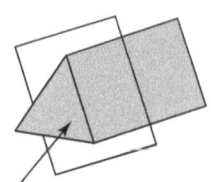

Equilateral triangle

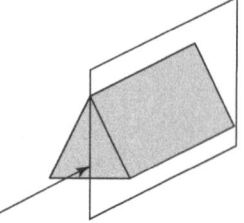

Equilateral triangle

Two triangular prisms with the same cross-section as the original.

Two triangular prisms with a right-angled triangle cross-section.

1   Describe two different ways in which each of these prisms can be cut in order to leave two identical prisms. Sketch your answers.

   a cuboid   **b** cylinder

2   **a**   Use the formula $v + f - e = 2$, where v = vertices, f = faces and e = edges, to copy and complete this table.

Name of solid	Number of vertices, v	Number of faces, f	Number of edges, e
tetrahedron	4	4	
cube	8		12
octahedron	6	8	
dodecahedron		12	30
icosahedron	12	20	

   b   Can you find out the collective name of these five solids?

Example

These are the plan, front elevation and side elevation for a solid.
Draw the solid.

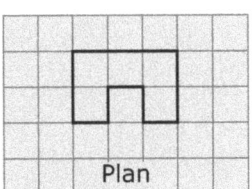
Plan

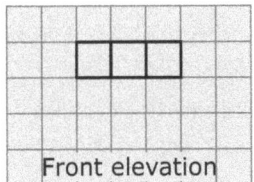
Front elevation

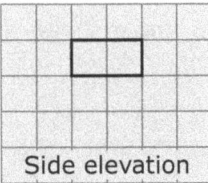
Side elevation

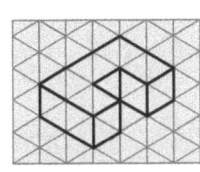

1 These solids are all made out of five one-centimetre cubes.
   For each solid draw **i** the plan **ii** the front elevation
   (from the direction of the arrow). Use square grid paper.

a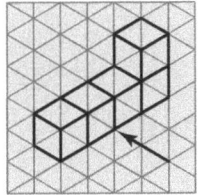

b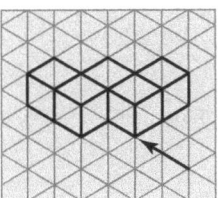

c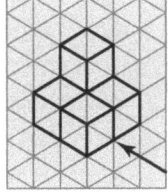

2 These are the plan, front elevation and side elevation for
   a solid. Draw the solid.

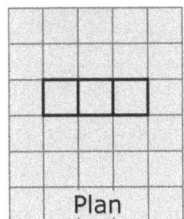
Plan

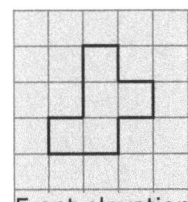
Front elevation

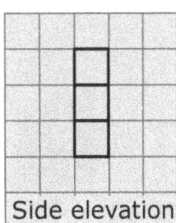
Side elevation

Example

Calculate **a** the volume **b** the surface area
of a 3cm by 6cm by 7cm cuboid.

- -

a Volume of a cuboid = length × width × height
$$= 3 \times 6 \times 7$$
$$= 126 \, cm^3$$

b Surface area = $(2 \times 3 \times 6) + (2 \times 3 \times 7) + (2 \times 6 \times 7)$
$$= 162 \, cm^2$$

1 Which of these nets will fold to form a cube?

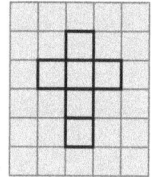 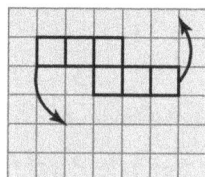 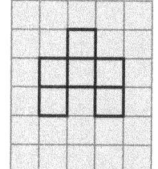 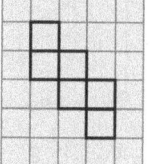

2 a Calculate the volume of this cuboid.
b Calculate the surface area of this cuboid.
c Draw a net of this cuboid.

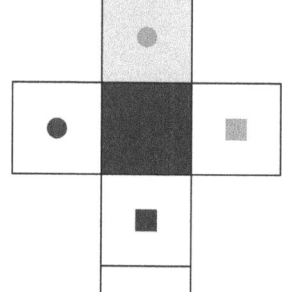

4 cm
3 cm
6 cm

3 Which of these cubes could be formed
from this net?

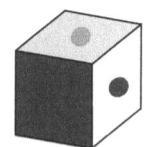

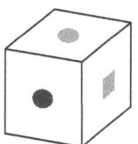

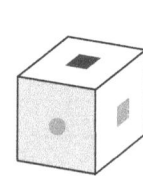

MyMaths.co.uk

1106 SEARCH

13a Sequences

This is a linear sequence.

3, ☐ , 17, 24, 31, ...

Find the missing value.

- -

The sequence increases in jumps of 7.

$(17 + 7 = 24, 24 + 7 = 31)$,

The missing value is $3 + 7 = 10$.

1 Copy each of these sequences and find the next two terms.

    a 3, 6, 9, 12, 15, ...
    b 4, 9, 14, 19, 24, ...

    c 50, 46, 42, 38, 34, ...
    d 26, 20, 14, 8, 2, ...

    e 3, 3.3, 3.6, 3.9, 4.2, ...
    f 1000, 100, 10, 1, 0.1, ...

2 Look at this number pattern.

$$2 \times 9 = 18$$
$$22 \times 9 = 198$$
$$222 \times 9 = 1998$$
$$2222 \times 9 = 19998$$

    a Write the next two lines in this pattern. Use your calculator to check your answers.

    b By relating the number of digits that are 2 in the question to the number of digits that are 9 in the answer, write the answer to $22\,222\,222 \times 9$.

3 Describe the rule for each sequence in words and find the missing values.

    a 2, 9, ☐ , 23, 30, ...
    b ☐ , 7, 11, 15, 19, ...

    c 2, $2\frac{1}{2}$, 3, ☐ , 4, ...
    d 7, ☐ , −3, ☐, −13, ...

Example

Find the position-to-term rule for the sequence
5, 14, 23, 32, 41, …
and use your rule to find the 100th term.

Each pair of terms has a difference of 9. This means that
the rule is 'Multiply by 9 and add or subtract something'.

Position	1	2	3	4	5
Multiply by 9	9	18	27	36	45
Subtract 4	5	14	23	32	41

So the position-to-term rule is 'Multiply by 9 and subtract 4'.
The 100th term is $100 \times 9 - 4 = 896$.

1 Generate the first five terms of the sequences described
by these position-to-term rules.
  a Add 1 to the position
  b Multiply the position by 5
  c Multiply by 3 and add 2
  d Divide by 3 and add 1
  e Multiply the position by itself and then by itself again.

2 For each sequence, complete this position-to-term rule 'Multiply
by ☐ and add ☐.'
  a 4, 7, 10, 13, 16, ….
  b 11, 13, 15, 17, 19, ….
  c 7, 12, 17, 22, 27, ….
  d $1\frac{1}{2}$, 2, $2\frac{1}{2}$, 3, $3\frac{1}{2}$, ….

3 Generate the first term and the 100th term of the sequences
defined by each position-to-term rule.
  a Position subtract 10
  b Multiply the position by 4 and add 3
  c Add 8 to the position and divide by 2
  d Multiply the position by itself and then by itself again.

4 Find a position-to-term rule for each of these linear sequences.
  a 3, 6, 9, 12, 15, …
  b 1, 4, 7, 10, 13, …

MyMaths.co.uk

Q 1173 SEARCH

Generate the first five terms of the sequence defined by
$T(n) = 5(n - 1)$

- -

$T(1) = 5 \times (1 - 1) = 0$
$T(2) = 5 \times (2 - 1) = 5$
$T(3) = 5 \times (3 - 1) = 10$
$T(4) = 5 \times (4 - 1) = 15$
$T(5) = 5 \times (5 - 1) = 20$

1 Write a sentence starting with 'I think of a number...' to describe
each of these expressions.

a $5n$ **b** $2n + 5$ **c** $\dfrac{n}{2}$

d $\dfrac{n}{3} + 1$ **e** $2(n + 7)$

2 Generate the first five terms of the sequences given
by these position-to-term rules where n is the position
and T is the term.

a $T(n) = n + 3$ **b** $T(n) = 8n$
c $T(n) = 2n + 5$ **d** $T(n) = 4(n - 1)$

3 The position-to-term rule of a sequence is
$T(n) = 10n + 5$. Find

a the first term ($n = 1$)   **b** $T(7)$
c the 10th term   **d** the 100th term.

4 Which of these sequences has the smallest 10th term?

n^2 $3(n + 4)$ $\dfrac{n}{2} + 1$ $7n - 5$

13d Finding a rule from a sequence

A linear sequence has third term 28 and fifth term 44.
What is the first term?

- -

Each consecutive pair of terms has the same difference.
There are two differences between the third and fifth terms.
44 − 28 = 16 so one difference is 8.
So the sequence is 12, 20, 28, 36, 44. The first term is 12.

1 Copy and complete the table to find the position-to-term rule in words
and in symbols for the sequence 8, 11, 14, 17, 20, ...

Position	1	2	3	4	5
Multiply by ☐	3	6	9		
Add ☐					
Term	8	11	14	17	20

2 Using the method in question **1**, find the position-to-term
formula for these linear sequences.
- **a** 4, 8, 12, 16, 20, ...
- **b** 10, 20, 30, 40, 50, ...
- **c** 3, 5, 7, 9, 11, ...
- **d** 5, 11, 17, 23, 29, ...

3 Choose one of these nth terms for each of these sequences.

n^2	$n + 3$	$5n - 2$	$10 - n$	$2n$	$4n + 3$

- **a** 4, 5, 6, 7, 8, ...
- **b** 2, 4, 6, 8, 10, ...
- **c** 7, 11, 15, 19, 23, ...
- **d** 3, 8, 13, 18, 23, ...
- **e** 9, 8, 7, 6, 5, ...
- **f** 1, 4, 9, 16, 25, ...

4 A linear sequence has second term 7 and fourth term 13.
- **a** What is the first term?
- **b** What is the 100th term?

MyMaths.co.uk

Q 1165, 1945   **SEARCH**

13e Sequences in context

Example

Fences are constructed with posts and crossbars.

Find a position-to-term rule that connects the number of posts (p) with the number of crossbars (C).

The number of crossbars increases by 3 each time because each pair of posts needs 3 crossbars to join them. The rule is 'Multiply the number of posts by 3 and subtract 3'.

Posts (p)	1	2	3	4	5
Multiply by 3 and subtract 3	0	3	6	9	12

So the position-to-term rule is $C = 3p - 3$.

1 Here is a pattern of dots.
   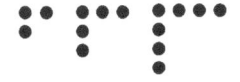
   a Write the number of dots needed in patterns four and five.
   b Copy and complete the table to find the position-to-term rule.

Position	1	2	3	4	5
Multiply by ☐ and add ☐					
Number of dots	3	5	7		

   c Explain why your formula works.

2 A shopkeeper stacks tins of food in these patterns.
   a Copy and complete the table.

Number of rows high (r)	1	2	3	4	5
Number of tins (T)	1	4			

   b Find a formula connecting the number of rows and number of tins.
   c Use your formula to work out the number of tins in a stack 20 rows high.

14a Mental methods of × and ÷ decimals

Example

Use the method of partitioning to work out
a 12 × 3.9   **b** 675 ÷ 9

- -

a 12 × 3.9 = 10 × 3.9 + 2 × 3.9 = 39 + 7.8 = 46.8

b 675 ÷ 9 = 630 ÷ 9 + 45 ÷ 9 = 70 + 5 = 75

1 Recall and use the methods of using factors, partitioning, compensation and doubling and halving to work these out mentally.
- **a** 12 × 50
- **b** 7.5 × 400
- **c** 11 × 6.4
- **d** 336 ÷ 8
- **e** 45 × 19
- **f** 23 × 21
- **g** 12 × 4.5
- **h** 3.8 × 5

2 You can think of multiplying by 0.1 and 0.01 as dividing by 10 and 100. Similarly, dividing by 0.1 and 0.01 can be thought of as multiplying by 10 and 100. Use these facts to calculate these questions mentally.
- **a** 14 × 0.1
- **b** 560 × 0.01
- **c** 1.6 ÷ 0.01
- **d** 250 × 0.1
- **e** 8.9 ÷ 0.1
- **f** 0.63 ÷ 0.01
- **g** 75 × 0.01
- **h** 0.09 ÷ 0.1

3 Without a calculator, calculate
- **a** 12^2
- **b** 4^3
- **c** 2^4
- **d** 1^3
- **e** $2^2 + 5^2$
- **f** $(2 + 5)^2$
- **g** $7^2 - 3^2$
- **h** $(7 - 3)^2$

4 Without a calculator, calculate
- **a** $\sqrt{25}$
- **b** $\sqrt{169}$
- **c** $\sqrt[3]{8}$
- **d** $\sqrt[3]{64}$
- **e** $\sqrt{9} + \sqrt{16}$
- **f** $\sqrt{9 + 16}$
- **g** $\sqrt{100} - \sqrt{64}$
- **h** $\sqrt{100 - 64}$

5 **a** Work these out mentally.
- **i** $2^2 × 3^2$
- **ii** $(2 × 3)^2$

   b Write what you notice.

6 **a** Work these out mentally.
- **i** $\sqrt{25} × \sqrt{4}$
- **ii** $\sqrt{25 × 4}$

   b Write what you notice.

Calculate 21.4 × 9.8 using the grid method.

First estimate the answer.   21.4 × 9.8 ≈ 20 × 10 = 200
Change the calculation to an equivalent whole number
calculation.   21.4 × 9.8 = 214 × 98 ÷ 100

×	200	10	4
90	200 × 90 = 18000	10 × 90 = 900	4 × 90 = 360
8	200 × 8 = 1600	10 × 8 = 80	4 × 8 = 32

18000 + 1600 + 900 + 80 + 360 + 32 = 20972
21.4 × 9.8 = 20972 ÷ 100 = 209.72

1 a Copy and complete this estimate.
629 × 18 ≈ 600 × ☐ = ☐
b Copy and complete this grid to calculate 629 × 18.

×	600	20	9
10	600 × 10 = 6000		
8			

2 Calculate
  a 26.1 × 5   **b** 42.8 × 9   **c** 52.5 × 6   **d** 8 × 71.4
  e 4.38 × 4   **f** 6.35 × 7   **g** 5 × 8.49   **h** 9 × 5.98

3 Calculate
  a 1.22 × 14   **b** 2.34 × 21   **c** 33 × 5.09   **d** 8.35 × 52
  e 27.5 × 3.4   **f** 4.5 × 4.86   **g** 5.19 × 7.3   **h** 9.2 × 84.7

4 a Shamoon tops up the tank of his car with 45 litres of diesel at a
cost of 97.4p per litre. Work out the total cost of the diesel.
  b Aysha buys 7.4 metres of voile at £8.95 per metre. Work out the
total cost of the voile.

14c Dividing decimals

Example

Calculate 67.7 ÷ 15 giving your answer to 2 decimal places.

Estimating the answer gives 67.7 ÷ 15 ≈ 70 ÷ 20 = 3.5

$$\begin{array}{r} 4.\ 5\ 1\ 3 \\ 15\overline{)67.^{7}7^{2}0^{5}0} \end{array}$$

Work out the answer to 3 decimal places in order to round to
2 decimal places. So 67.7 ÷ 15 = 4.51 (2dp)

1 Calculate these using an appropriate method. The whole number
answers can all be found in the box.

| 33 29 48 57 41 52 31 86 |

a 246 ÷ 6   **b** 416 ÷ 8   **c** 774 ÷ 9
d 336 ÷ 7   **e** 434 ÷ 14   **f** 348 ÷ 12
g 924 ÷ 28   **h** 1368 ÷ 24

2 Calculate these. Where appropriate, give your answer as a decimal
to 1 decimal place.

a 128 ÷ 6   **b** 194 ÷ 9   **c** 164 ÷ 7
d 233 ÷ 11   **e** 97.7 ÷ 15   **f** 86.9 ÷ 17
g 80.8 ÷ 18   **h** 95.6 ÷ 19

3 a A 750 ml bottle of Softy fabric conditioner claims to last for
21 washes. How much Softy should be used for one wash?
Give your answer to 2 decimal places.
b Softy fabric conditioner costs £1.38.
How much does each wash cost to the nearest penny?
c Comfy fabric conditioner is sold in 500 ml bottles for the same price.
One 30 ml capful of Comfy should be used per wash. Which fabric
conditioner is the best value for money?

14d Calculator methods 3

Flora was born 239 days ago.

Calculate how old Flora is in weeks and days.

- -

Divide by 7 (7 days in a week) $239 ÷ 7 = 34.14285714...$

Convert the remainder $0.14285714... × 7 = 1$
by multiplying by the divisor.

So Flora is 34 weeks and 1 day old.

1   Work out the age of these babies in weeks and days.
   a Holly 96 days
   b Lily 153 days
   c Annabelle 304 days

2   Convert these measurements of time to the units given in brackets.
   a 3.4 hours (hours and minutes)   **b** 136 minutes (hours and minutes)
   c 2.125 days (days and hours)   **d** 525 hours (days and hours)
   e 6.3 minutes (mins and secs)   **f** 3280 seconds (mins and secs)
   g 11.8 years (years and days)   **h** 4500 days (years and days)

3   Give your answers to these questions in a form appropriate to
   the situation.
   a Xavier is flying to Switzerland to sell ski-goggles. One boxed
        pair of goggles has a mass of 300g. Xavier must not exceed his
        baggage allowance of 23kg. How many pairs of goggles can he
        take and how much baggage allowance is left over?
   b Xavier buys each pair of ski-goggles for €188.50 and plans to
        sell them for €220. If he sells 45 pairs on his trip, how much
        profit will he make?

Example

Harry the cat has 24 pouches of food. 9 of these pouches are tuna and the rest are chicken. What proportion of the pouches are chicken? Give your answer as

a a fraction **b** a percentage.

- -

$24 - 9 = 15$ so 15 of the pouches are chicken.
The proportion that are chicken is

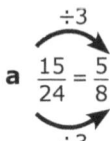

a $\dfrac{15}{24} = \dfrac{5}{8}$　　　　**b** $(5 \div 8) \times 100\% = 62.5\%$

1 Write each of these fractions as hundredths. Then convert each fraction to
i a decimal **ii** a percentage.

a $\dfrac{3}{4}$ **b** $\dfrac{2}{5}$ **c** $\dfrac{11}{20}$ **d** $\dfrac{26}{25}$

2 Use a calculator to change each of these fractions to
i a decimal **ii** a percentage.

a $\dfrac{1}{8}$ **b** $\dfrac{19}{40}$ **c** $\dfrac{8}{9}$ **d** $\dfrac{7}{32}$

3 Michael has a haulage business with 16 lorries. Five of his lorries are white and all the rest are blue. Six of his lorries are in need of repair.
a Write the proportion of lorries that are in need of repair as
i a fraction in its simplest form
ii a percentage.
b What proportion of Michael's lorries are blue?

4 Lynda has a library of books. $\dfrac{3}{4}$ of her books are novels and $\dfrac{3}{16}$ of her books are reference books. The rest of her books are on the subject of Italy.
a If Lynda has 200 books, how many are novels?
b What proportion of Lynda's books are on the subject of Italy?

MyMaths.co.uk

Q 1015 SEARCH

Example

8 jam tarts cost £1.20. Work out the cost of 20 jam tarts.

- -

The cost of 1 jam tart is £1.20 ÷ 8 = £0.15
The cost of 20 jam tarts is 20 × £0.15 = £3.00

1 Copy and complete this table to work out the cost of 5 metres of fabric
if 12 metres of fabric costs £546.

Amount of fabric in metres	Cost in £
÷ 12 ⟨ 12 → 1 ⟩ ×5 ⟨ → 5	÷ 12 ⟨ 546 →

2 40 litres of diesel costs £38.20. Work out the cost of
   a 1 litre of diesel   **b** 5 litres of diesel
   c 18 litres of diesel   **d** 45 litres of diesel.

3 This is a recipe for making 8 scones.
Write out a recipe for making
12 scones.

	200 g	self-raising flour
	50 g	soft margarine
	25 g	sugar
	50 g	currants
	1	egg

4 Nappies come in two pack sizes.
   a Is the number of nappies in
      a pack in direct proportion
      with the cost?
   b Which pack is better value
      for money? Show your
      working.

Cost £5.92   Cost £8.96

In 1982, the ratio of sheep to people in New Zealand was $20:1$.
If there were 70.3 million sheep, how large was the population?

70.3 million is 20 parts.
1 part is 70.3 million $\div$ 20 = 3.515 million.
The population was approximately 3.5 million people.

1   Write each of these ratios in its simplest form.
   a $2:10$ **b** $12:8$
   c $15:25$ **d** $6:2:10$
   e   £18:£6   **f** $6\,kg:15\,kg$
   g $250\,g:5\,kg$ **h** $24p:£1.80$
   i   2.5 hours : 100 minutes   **j** $56\,cm:2.8\,m:140\,mm$

2   In a class of 24 students there are 18 boys.
   a   Write the ratio of boys to girls in its simplest form.
   b   Write the ratio of girls to boys in its simplest form.

3   The ratio of Dora's pocket money to Diego's pocket money is $4:5$.
   a   If Dora receives £4.80 per week, work out how much pocket
            money Diego receives.
   b   Dora has her weekly pocket money increased by 45p. Work out
            the new ratio of Dora's pocket money to Diego's pocket money.

4   A map has a scale of $1:20\,000$.
   a   What is the distance in real life of a measurement of 8 cm on the
            map? Give your answer in kilometres.
   b   What is the measurement on the map of a distance of 300 m
            in real life? Give your answer in centimetres.

5   The angles in a triangle are in the ratio $2:2:5$.
   a   Work out the sizes of the angles of the triangle.
   b   What type of triangle is this?

James is making a glass of his favourite cordial. The bottle suggests using water and cordial in the ratio 4 : 1. James's glass has a capacity of 250 ml. How much water should he pour into his glass?

Divide 250 ml into 1 + 4 = 5 equal parts.
1 part is 250 ml ÷ 5 = 50 ml.
James requires 4 parts of water so 4 × 50 ml = 200 ml of water.

1 Divide each of these quantities in the ratio given in brackets.

   a £20 (1 : 4)   **b** 50 cm (3 : 7)

   c 55 kg (3 : 8)   **d** 108 litres (3 : 4 : 5)

2 An alloy is a mixture of two elements, at least one of which is a metal.

   a Bronze is made from combining copper and tin in the ratio 22 : 3. Calculate, in kilograms, the mass of copper and tin in a bronze statue of mass 100 kg.

   b Brass is made from combining copper and zinc in the ratio 9 : 1. Calculate, in grams, the mass of copper and zinc in a brass sundial of mass 4 kg.

3 In the United Kingdom, silver-coloured coins are made from cupro-nickel, an alloy of copper and nickel. In cupro-nickel, the ratio of copper to nickel is 3 : 1.

   a A 50p coin has a mass of 8 g. Calculate the mass of copper and nickel in a 50p coin.

   b A 10p coin has a mass of 6.5 g. Calculate the mass of copper and nickel in a 10p coin.

4 Annabelle is making a fruity drink. She uses apple juice, ginger ale and lemonade in the ratio 4 : 2 : 1. Annabelle's glass has a capacity of 280 ml. Work out the amount of each ingredient that she needs.

Example

Angus and Hamish share £40 in the ratio $3:5$. Write the proportion of money that Hamish receives and use this to calculate the amount.

There are $3 + 5 = 8$ parts and so Hamish receives $\frac{5}{8}$ of the money.
$\frac{5}{8}$ of £40 $= \frac{1}{8} \times 5 \times 40 = \frac{200}{8} =$ £25. So Hamish receives £25.

1 In a class of 25 students, 3 are left-handed.
   a Write the ratio of left-handed students to right-handed students.
   b Write the proportion of students that are left-handed.
   c The proportion of left-handed students in this class is the same as in the UK population. If the UK population is approximately 60 million people, estimate the number of left-handed people.

2 Jack makes a 5 litre tin of pale pink paint by mixing 3.5 litres of brilliant white paint with 1.5 litres of ruby red paint.
   a Write the ratio of ruby red to brilliant white paint.
   b Write the proportion of the paint that is ruby red.

3 To make chocolate buttercream for a batch of fairy cakes, Florence mixes together 180 g of icing sugar, 120 g of butter and 60 g of cocoa powder.
   a Write the ratio of icing sugar to butter to cocoa powder.
   b What proportion of the buttercream is cocoa powder?
   c Scale up this recipe to make 480 g of buttercream.

4 In the UK there are currently some 2 million North American grey squirrels and only 160 000 native red squirrels.
   a Write the ratio of grey to red squirrels in its simplest form.
   b What proportion of the UK squirrel population is red?
   c Of the red squirrels, $\frac{3}{16}$ live in England. Write the ratio of red squirrels in England to red squirrels in the rest of the UK.

MyMaths.co.uk

Q 1039   SEARCH

Example

A pair of designer sunglasses costs £135. In an end-of-season sale the sunglasses are reduced by 5%. Work out the cost of the sunglasses in the sale.

$(100 - 5)$% of the old price = 95% of £135
$$= 0.95 \times 135 = £128.25$$

1 Work out each answer and then write the letter that is in that position in the alphabet. For example, if your answer is 8 write h as h is in the 8th position in the alphabet. Find the hidden word.

a 20% of 95   b $17\frac{1}{2}$% of 120   c 8% of 200   d 12.5% of 40

e 45% of 40   f 1% of 200   g $33\frac{1}{3}$% of 36   h 4% of 625

2 Work out these percentages using a calculator.

a Girls aged 11−14 require 1800 calories per day. Molly eats an Easter egg and immediately consumes 9.5% of her daily calories. How many calories are there in the Easter egg?

b Johanna spends £24.40 on her meal at a French bistro. A 12.5% service charge is added to her bill. How much is the service charge?

c A bank offers an account with an interest rate of 6.5%. John invests £3000. How much interest will he earn in one year?

3 Calculate these percentage changes.

a Increase £15 by 10%   b Increase 24 m by 25%

c Decrease 50 kg by 12%   d Increase 19 litres by 6%

4 a A 500 ml bottle of surface cleaner is advertised with an additional 20% extra free. How much surface cleaner does the promotional bottle contain?

b House prices in the North West increased by 12% from 2005 to 2006. In 2005 the average house price was £138 000. Calculate the average house price in 2006.

c A car bought for £14 300 in 2005 depreciates by 22% in the first year. What is this car worth in 2006?

MyMaths.co.uk

Example

P(X) = 0.4. Work out P(not X).

$$P(\text{not } X) = 1 - P(X)$$
$$= 1 - 0.4$$
$$= 0.6$$

1

```
|----------|----------|----------|----------|----------|
0                                                      1
```

Copy this probability scale and place these cards at the correct points on the scale.

| Evens | Unlikely | Certain | Impossible | Likely |

2 Show the likelihood of each of these events happening on a probability scale.

a If you toss a coin you will get a 'head'.

b If you throw a dice you will get a six.

c Tomorrow will be Tuesday.

d It will snow tomorrow.

e When you next write your first name, you will write the letter 'a'.

f Your maths teacher will win the National Lottery this week.

g Your head teacher will grant your maths class an extra week's summer holiday for exceptional work.

3 | P(A) = 0.7 | P(B) = 0 | P(C) = 0.5 | P(D) = 0.2 |

a Draw a probability scale and mark these events on it.

b What is the probability that event A does not happen?

c Write P(not D).

d Describe P(B) in words.

e Give an example of an event that has the same probability as event C.

MyMaths.co.uk

Q 1209 **SEARCH**

Four discs numbered 2, 4, 6 and 8 are placed in a bag. Charlie chooses a disc at random. What is the probability that he chooses

a a 6   **b** an even number   **c** a number more than 4?

---

a $P(6) = \frac{1}{4}$ There is one 6 amongst four numbers.

b $P(\text{an even number}) = \frac{4}{4} = 1$ All four numbers are even numbers.

c $P(\text{a number more than 4}) = \frac{2}{4} = \frac{1}{2}$ 6 and 8 are more than 4.

1 You spin this spinner once.

   a List all the possible outcomes.

   b Are all the outcomes equally likely? Explain your answer.

   c Write

     i P(the spinner lands on red)

     ii P(the spinner lands on blue).

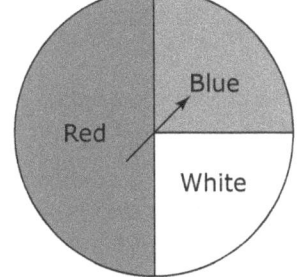

2 You throw a fair dice.

   a List all the possible outcomes.

   b Are all the outcomes equally likely? Explain your answer.

   c Write

     i P(the dice shows 3)

     ii P(the dice shows an even number).

3 A tin contains 10 metallic crayons. Five crayons are gold, three crayons are silver and two crayons are bronze. Freya chooses a crayon at random. What is the probability that the crayon is

   a silver   **b** gold   **c** not bronze   **d** pewter?

4 Eight balls numbered 1 to 8 are placed in a hat. Amber selects a ball at random. What is the probability that she selects

   a a 4

   b an odd number

   c a number less than 7

   d a rabbit?

Example

John throws an ordinary dice.
a Are the events 'John throws a 6' and 'John throws an odd number' mutually exclusive?
b What is the probability that John throws a 6 or an odd number?

- -

a Yes, because these events cannot occur together. A 6 is not odd.
b P(6 or an odd number) = P(6) + P(odd number)

$$= \frac{1}{6} + \frac{3}{6} = \frac{4}{6} = \frac{2}{3}$$

1 When a coin is tossed the events 'toss a head' and 'toss a tail' are mutually exclusive. True or false?

2 Jack is playing with his trolley of bricks. The bricks are numbered 1, 2, 3, 4 and 5. The table gives the probability that Jack selects a brick painted with a particular number. What is the probability that Jack selects a brick

a numbered 5
b numbered 3 or 4
c numbered 1, 2 or 3?

	P(1) = 0.3
	P(2) = 0.1
	P(3) = 0.25
	P(4) = 0.15
	P(5) =

3 A multi-pack bag of crisps contains 8 packets. Three of the packets are flavoured with sea salt, three with balsamic vinegar and the rest with cracked black pepper. Ashim chooses a packet of crisps at random.

a Are the events 'choose a packet flavoured with sea salt', 'choose a packet flavoured with balsamic vinegar' and 'choose a packet flavoured with cracked black pepper' mutually exclusive?
b What is the probability that Ashim chooses a packet flavoured with cracked black pepper or a packet flavoured with sea salt?

MyMaths.co.uk

Q 1263 SEARCH

Paula tosses a coin 100 times and obtains these results.

a Estimate the probability that on the next toss, Paula will get
  i a head   ii a tail.

| Head | 45 times |
| Tail | 55 times |

b How could Paula improve her probability estimates?

- -

a i P(Head) = $\frac{45}{100} = \frac{9}{20}$ ii P(Tail) = $\frac{55}{100} = \frac{11}{20}$

b She should toss the coin more times – the more trials conducted, the better the probability estimate.

---

1 Jason has a bag of discs. He selects one disc at a time without looking, records the symbol on the disc and then replaces the disc in the bag. Here are his results.

Symbol	Star	Circle	Heart	Question mark
No. of times drawn	24	30	18	28

  a Work out the number of times Jason selected a disc.
  b Jason selects another disc. Estimate the probability of selecting
    i a circle ii a question mark iii a star or a heart.
  c The bag only contains discs which are printed with one of four symbols. These are stars, circles, hearts and question marks. True or false? Explain your answer.

2 Luke has a pencil case containing blue and black pens. He chooses a pen at random, records its colour and replaces the pen in his case. Here are his results.
  a Estimate the probability that the next pen Luke selects is i blue ii black.
  b Luke actually has 6 pens in his case. Estimate how many of each colour he has.

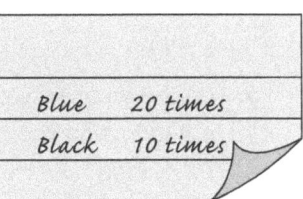

| Blue | 20 times |
| Black | 10 times |

Example

A dice in the shape of a tetrahedron has faces numbered from 1 to 4. The dice is thrown 100 times. How many times would you expect to see

a a 1   **b** an odd number?

a $P(1) = \frac{1}{4}$ and $\frac{1}{4} \times 100 = 25$ times

b P(an odd number) $= \frac{2}{4} = \frac{1}{2}$ and $\frac{1}{2} \times 100 = 50$ times.

1 A dice is thrown 120 times.

   a How many times would you expect to see

   i a 6   **ii** a 1 or a 2   **iii** an even number?

Here are the results of these 120 throws.

Score	1	2	3	4	5	6
Frequency	18	15	45	14	20	8

   b Is the dice biased? Explain your answer.

2 A dice is biased so that the probability of throwing a 6 is 0.35.

   a After 100 throws of the dice, how many sixes would you expect?

   b After 500 throws of the dice, how many sixes would you expect?

3 Simon, Robbie and Flip all test the same 10p coin for bias. These are their results.

Simon		Robbie		Flip	
20 tosses		50 tosses		200 tosses	
H	T	H	T	H	T
15	5	32	18	112	88

   a Giving your answer as a decimal, estimate the probability of tossing a head using the results table for

     i Simon   **ii** Robbie   **iii** Flip.

   b What conclusion would you draw from each person's data?

   c Which conclusion is likely to be the most reliable? Why?

   d What should Simon, Robbie and Flip do with their data to reach an even more reliable conclusion?

MyMaths.co.uk

Q 1263   SEARCH

a Use a Venn diagram to sort the letters of the word 'FRACTIONS' into:
V = {the set of vowels} and W = {the letters in the word 'MATHS'}.
Use numbers to represent the number of letters in each set.

b How many letters are in the union of V and W?

---

a V = {A I O}   3 letters
W = {A T S}   3 letters
V ∩ W = {A}   1 letter
Remaining letters: F R C N

b 2 + 1 + 2 = 5 letters

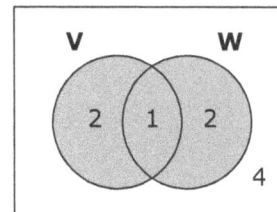

1 A warehouse uses a four-digit code   X32C   Y23A   X11A   Z23B
for each item it stores.   X12B   Z33A   Y13B   X21C

The warehouse manager picks eight codes.
He sorts them into A = {codes starting with X} and
B = {codes containing 2}.

a List the codes in the intersection of A and B.

b Sort the codes in a Venn diagram. Use numbers to represent the
number of codes in each set.

c How many codes are in A union B?

d How many codes are in A'?

2 Katya asks 20 students at her school if they
are playing hockey or tennis in PE this term.
Her results are shown on the Venn diagram.
How many students are

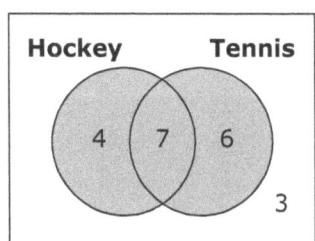

a in the intersection of hockey and tennis

b playing hockey

c not playing tennis

d in the union of hockey and tennis?

e Describe the shaded region in words.

f What fraction of students are playing either hockey or tennis,
but not both?

🌐 **MyMaths**.co.uk
🔍 1921   **SEARCH**
Statistics and probability Probability   **99**

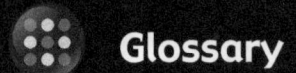

add, addition (+)	Addition is the sum of two numbers or quantities.
adjacent (side)	Adjacent sides are next to each other and are joined by a common vertex.
algebra	Algebra is the branch of mathematics where symbols or letters are used to represent numbers.
amount	Amount means total.
angle: acute, obtuse, right, reflex	An angle is formed when two straight lines cross or meet each other at a point. The size of an angle is measured by the amount one line has been turned in relation to the other.

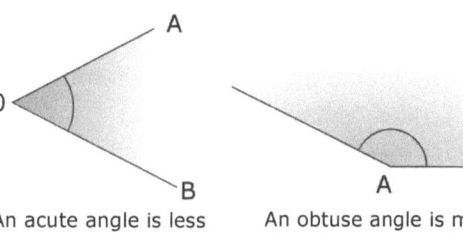

An acute angle is less than 90°

An obtuse angle is more than 90° but less than 180°

A right angle is a quarter of a turn, or 90°

A reflex angle is more than 180° but less than 360°

angles at a point   Angles at a point add up to 360°.

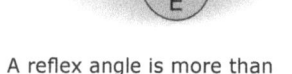

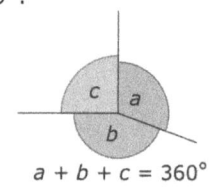

$a + b + c = 360°$

angles on a straight line	Angles on a straight line add up to 180°.
	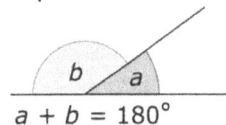
	$a + b = 180°$
approximate, approximately	An approximate value is a value that is close to the actual value of a number.
approximately equal to ($\approx$)	Approximately equal to means almost the same size.
area	The area of a surface is a measure of its size. The units for area are square millimetre, square centimetre, square metre, square kilometre and so on.
average	An average is a representative value of a set of data.
axis, axes	An axis is one of the lines used to locate a point in a coordinate system.
bar chart	A bar chart is a diagram that uses rectangles of equal width to display data. The frequency is given by the height of the rectangle.
bar-line graph	A bar-line graph is a diagram that uses lines to display data. The lengths of the lines are proportional to the frequencies.
base	The lower horizontal edge of a shape is usually called the base. Similarly, the base of a solid is its lower face.
between	Between means in the space bounded by two limits.
brackets	Operations within brackets should be carried out first.

calculate, calculation	Calculate means work out using a mathematical procedure.
calculator	You can use a calculator to perform calculations.
cancel, cancellation	A fraction is cancelled down by dividing the numerator and denominator by a common factor.

For example, $\dfrac{24}{40} = \dfrac{3}{5}$ (÷ 8)

capacity: litre	Capacity is a measure of the amount of liquid a 3D shape will hold.
centre of rotation	The centre of rotation is the fixed point about which a rotation takes place.

centre of rotation

certain	An event that is certain will definitely happen.
chance	Chance is the probability of something happening.
class interval	A class interval is a group that you put data into to make it easier to handle.
common factor	A common factor is a factor of two or more numbers. For example, 2 is a common factor of 4 and 10.
compare	Compare means to assess the similarity of.
congruent	Congruent shapes are exactly the same shape and size.
consecutive	Consecutive means following on in order. For example 2, 3 and 4 are consecutive integers.
construct	To construct means to draw a line, angle or shape accurately.

continue	Continue means carry on.
convert	Convert means to change.
coordinate pair	A coordinate pair is a pair of numbers that give the position of a point on a coordinate grid. For example, (3, 2) means 3 units across and 2 units up.
coordinate point	A coordinate point is the point described by a coordinate pair.
coordinates	Coordinates are the numbers that make up a coordinate pair.
data	Data are pieces of information.
data collection sheet	A data collection sheet is a sheet used to collect data. It is sometimes a list of questions with tick boxes for collecting answers.
decimal number	A decimal number is a number written using a decimal point.
decimal place (dp)	Each column after the decimal point is called a decimal place. For example, 0.65 has two decimal places (2dp).
degree (°)	A degree is a measure of turn. There are 360° in a full turn.
denominator	The denominator is the bottom number in a fraction. It shows how many parts there are in total.
difference	You find the difference between two amounts by subtracting one from the other.
digit	A digit is any of the numbers 0, 1, 2, 3, 4, 5, 6, 7, 8, 9.

direction	The direction is the orientation of a line in space.
distance	The distance between two points is the length of the line that joins them.
divide, division (÷)	Divide means share equally.
divisible, divisibility	A whole number is divisible by another if there is no remainder left.
divisor	The divisor is the number that does the dividing. For example, in 14 ÷ 2 = 7 the divisor is 2.
double, halve	Double means multiply by two. Halve means divide by two.
edge (of solid)	An edge is a line along which two faces of a solid meet.

edge

equal (sides, angles)	Equal sides are the same length. Equal angles are the same size.
equally likely	Events are equally likely if they have the same probability.
equals (=)	Equals means having exactly the same value or size.
equation	An equation is a statement using an = sign to link two expressions.
equivalent, equivalence	Equivalent fractions are fractions with the same value.
estimate	An estimate is an approximate answer.
evaluate	Evaluate means find the value of an expression.
exact, exactly	Exact means completely accurate. For example, 3 divides into 6 exactly.

experiment	An experiment is a test or investigation to gather evidence for or against a theory.
expression	An expression is a collection of numbers and symbols linked by operations but not including an equals sign.
face	A face is a flat surface of a solid.

face

factor	A factor is a number that divides exactly into another number. For example, 3 and 7 are factors of 21.
fair	In a fair experiment there is no bias towards any particular outcome.
fraction	A fraction is a way of describing a part of a whole.

$\frac{2}{5}$

frequency	Frequency is the number of times something occurs.
frequency diagram	A frequency diagram uses bars to display grouped data. The height of each bar gives the frequency of the group, and there is no space between the bars.
function	A function is a rule. For example, +2, −3, ×4 and ÷5 are all functions.
function machine	A function machine links an input value to an output value by performing a function.
generalise	Generalise means formulate a general statement or rule.
generate	Generate means produce.
graph	A graph is a diagram that shows a relationship between variables.

greater than (>)	Greater than means more than.
	For example 4 > 3.
grid	A grid is used as a background to plot coordinate points. It is usually square.
height	Height is the vertical distance from the base to the top of a shape.
highest common factor (HCF)	The highest common factor is the largest factor that is common to two or more numbers.
	For example, the HCF of 12 and 8 is 4.
horizontal	Horizontal means flat and level with the ground.
hundredth	A hundredth is 1 out of 100.
	For example 0.05 has 5 hundredths.
impossible	An event is impossible if it definitely cannot happen.
improper fraction	An improper fraction is a fraction where the numerator is greater than the denominator. For example, $\frac{8}{5}$.
increase, decrease	Increase means make greater. Decrease means make less.
input, output	Input is data fed into a machine or process. Output is the data produced by a machine or process.
integer	An integer is a positive or negative whole number (including zero). The integers are: ..., -3, -2, -1, 0, 1, 2, 3, ...
interpret	You interpret data whenever you make sense of it.

intersect, intersection	Two lines intersect at the point, or points, where they cross.
	intersection
interval	An interval is the size of a class or group in a frequency table.
inverse	An inverse operation has the opposite effect to the original operation. For example, multiplication is the inverse of division.
label	A label is a description of a diagram or object.
length: millimetre, centimetre, metre, kilometre; mile, foot, inch	Length is a measure of distance. It is often used to describe one dimension of a shape.
less than (<)	Less than means smaller than. For example, 3 is less than 4 or 3 < 4.
likelihood	Likelihood is the probability of an event happening.
likely	An event is likely if it will happen more often than not.
line	A line joins two points.
line of symmetry	A line of symmetry is a line about which a 2D shape can be folded so that one half of the shape fits exactly on the other half.
line symmetry	A shape has line symmetry if it has a line of symmetry.
lowest common multiple (LCM)	The lowest common multiple is the smallest multiple that is common to two or more numbers. For example the LCM of 4 and 6 is 12.
lowest terms	A fraction is in its lowest terms when the numerator and denominator have no common factors.

mapping

A mapping is a rule that can be applied to a set of numbers to give another set of numbers.

mass: gram, kilogram; ounce, pound

Mass is a measure of the weight of an object.

mean

The mean is an average value found by adding all the data values and dividing by the number of pieces of data.

measure

When you measure something you find the size of it.

median

The median is the average value which is the middle value when the data is arranged in size order.

mirror line

A mirror line is a line of symmetry.

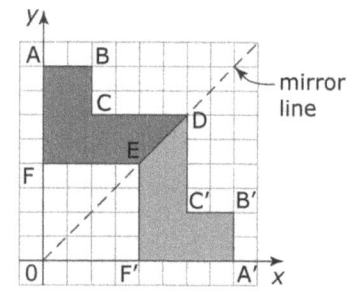

mixed number

A mixed number has a whole number part and a fraction part.

For example, $3\frac{1}{2}$ is a mixed number.

modal class

The modal class is the most commonly occurring class when the data is grouped. It is the class with the highest frequency.

mode

The mode is an average which is the data value that occurs most often.

multiple

A multiple of an integer is the product of that integer and any other.

For example, these are multiples of 6: $6 \times 4 = 24$ and $6 \times 12 = 72$.

multiply, multiplication (×)	Multiplication is the operation of combining two numbers or quantities to form a product.
nearest	Nearest means the closest value.
negative	A negative number is a number less than zero.
net	A net is a 2D arrangement that can be folded to form a solid shape.
***n*th term**	The *n*th term is the general term of a sequence.
numerator	The numerator is the top number in a fraction. It shows how many parts you are dealing with.
object, image	The object is the original shape before a transformation. An image is the same shape after a transformation.
operation	An operation is a rule for processing numbers or objects. The basic operations are addition, subtraction, multiplication and division.
opposite (sides, angles)	Opposite means across from.
order	To order means to arrange according to size or importance.
order of operations	The conventional order of operations is BIDMAS: brackets first, then indices, then division and multiplication, then addition and subtraction.
order of rotation symmetry	The order of rotation symmetry is the number of times that a shape will fit on to itself during a full turn.
origin	The origin is the point where the x- and y-axes cross, that is $(0, 0)$.
outcome	An outcome is the result of a trial or experiment.

parallel	Two lines that always stay the same distance apart are parallel. Parallel lines never cross or meet.
partition; part	To partition means to split a number into smaller amounts, or parts. For example, 57 could be split into 50 + 7, or 40 + 17.
percentage (%)	A percentage is a fraction expressed as a number of parts per hundred.
perimeter	The perimeter of a shape is the distance around it. It is the total length of the edges.
perpendicular	Two lines are perpendicular to each other if they meet at a right angle.
pie chart	A pie chart uses a circle to display data. The angle at the centre of a sector is proportional to the frequency.
place value	The place value is the value of a digit in a decimal number. For example, in 3.65 the digit 6 has a value of 6 tenths.
polygon: pentagon, hexagon, octagon	A polygon is a closed shape with three or more straight edges.

A pentagon has five sides.   A hexagon has six sides.   An octagon has eight sides.

positive	A positive number is greater than zero.
predict	Predict means forecast in advance.
prime	A prime number is a number that has exactly two different factors, itself and 1.

probability	Probability is a measure of how likely an event is.
probability scale	A probability scale is a line numbered 0 to 1 or 0% to 100% on which you place an event based on its probability.
product	The product is the result of a multiplication.
proportion	Proportion compares the size of a part to the size of a whole. You can express a proportion as a fraction, decimal or percentage.
protractor (angle measurer)	A protractor is an instrument for measuring angles in degrees.
quadrant	A coordinate grid is divided into four quadrants by the x- and y-axes.

quadrilateral: arrowhead, kite, parallelogram, rectangle, rhombus, square, trapezium

A quadrilateral is a polygon with four sides.

rectangle   parallelogram   kite

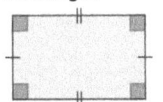 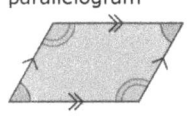 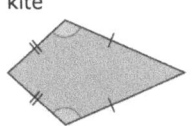

All angles are right angles.   Two pairs of parallel sides.   Two pairs of adjacent sides equal.

rhombus   square   trapezium

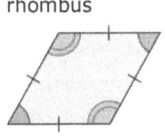 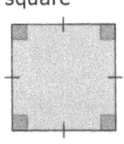 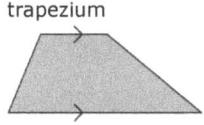

All sides the same length. Opposite angles equal.   All sides and angles equal.   One pair of parallel sides.

questionnaire	A questionnaire is a list of questions used to gather information in a survey.
quotient	A quotient is the result of a division.

random	A selection is random if each object or number is equally likely to be chosen.
range	The range is the difference between the largest and smallest values in a set of data.
ratio	Ratio compares the size of one part with the size of another part.
reflect, reflection	A reflection is a transformation in which corresponding points in the object and the image are the same distance from the mirror line. 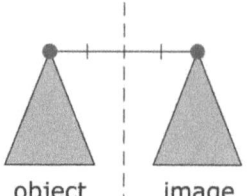
reflection symmetry	A shape has reflection symmetry if it has a line of symmetry.
regular	A regular polygon has equal sides and equal angles.
relationship	A relationship is a link between objects or numbers.
remainder	A remainder is the amount left over when one quantity is divided by another. For example, 9 ÷ 4 = 2 remainder 1 or 2 r 1.
represent	You represent data whenever you display it in the form of a diagram.
rotate, rotation	A rotation is a transformation in which every point in the object turns through the same angle relative to a fixed point.
rotation symmetry	A shape has rotation symmetry if when turned it fits onto itself more than once during a full turn.

round	You round a number by expressing it to a given degree of accuracy. For example, 639 is 600 to the nearest 100 and 640 to the nearest 10. To round to one decimal place means to round to the nearest tenth. For example 12.47 is 12.5 to 1dp.
rule	A rule describes the link between objects or numbers. For example, the rule linking 2 and 6 may be +4 or ×3.
ruler	A ruler is an instrument for measuring lengths.
sequence	A sequence is a set of numbers or objects that follow a rule.
shape	A shape is made by a line or lines drawn on a surface, or by putting surfaces together.
side (of 2D shape)	A side is a line segment joining vertices.
sign	A sign is a symbol used to denote an operation.
simplest form	A fraction (or ratio) is in its simplest form when the numerator and denominator (or parts of the ratio) have no common factors. For example, $\frac{3}{5}$ is expressed in its simplest form.
simplify	To simplify an expression you gather all like terms together into a single term.
sketch	A sketch shows the general shape of a graph or diagram.
solid (3D) shape: cube, cuboid, prism, pyramid,	A solid is a shape formed in three-dimensional space.

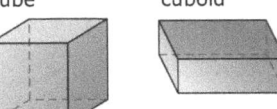

cube   cuboid   prism

six square faces six rectangular faces the cross section is constant

solution (of an equation) The solution of an equation is the value of the variable that makes the equation true.

solve (an equation) To solve an equation you need to find the value of the variable that will make the equation true.

spin, spinner A spinner is an instrument for creating random outcomes, usually in probability experiments.

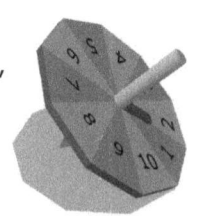

square-based pyramid, tetrahedron

pyramid

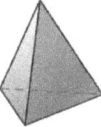

the faces meet at a common vertex angles equal

tetrahedron

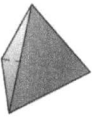

all faces are equilateral triangles

square-based pyramid

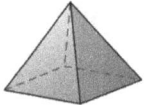

the base is a square

square number, squared If you multiply a number by itself the result is a square number.

For example 25 is a square number because $5^2 = 5 \times 5 = 25$.

square root A square root is a number that when multiplied by itself is equal to a given number.

For example, $\sqrt{25} = 5$, because $5 \times 5 = 25$.

statistic, statistics Statistics is the collection, display and analysis of information.

straight-line graph When coordinate points lie in a straight line they form a straight-line graph. It is the graph of a linear equation.

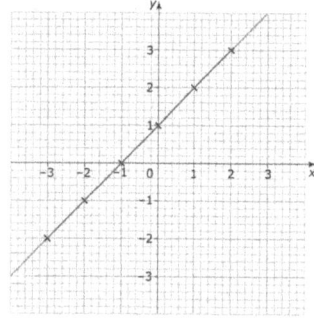

substitute	When you substitute you replace part of an expression with its value.
subtract, subtraction (−)	Subtraction is the operation that finds the difference in size between two numbers.
sum	The sum is the total and is the result of an addition.
surface, surface area	The surface area of a solid is the total area of its faces.
survey	A survey is an investigation to find information.
symbol	A symbol is a letter, number or other mark that represents a number or an operation.
symmetrical	A shape is symmetrical if it is unchanged after a rotation or reflection.
table	A table is an arrangement of information, numbers or letters usually in rows and columns.
tally	You use a tally mark to represent an object when you collect data. Tally marks are made in groups of five to make it easier to count them.
temperature: degrees Celsius, degrees Fahrenheit	Temperature is a measure of how hot something is.
tenth	A tenth is 1 out of 10 or $\frac{1}{10}$. For example 0.5 has 5 tenths.
term	A term is a number or object in a sequence. It is also part of an expression.
thousandth	A thousandth is 1 out of 1000 or $\frac{1}{1000}$. For example, 0.002 has 2 thousandths.

three-dimensional (3D)	Any solid shape is three-dimensional.
total	The total is the result of an addition.
transformation	A transformation moves a shape from one place to another.
translate, translation	A translation is a transformation in which every point in an object moves the same distance and direction. It is a sliding movement.
triangle: equilateral, isosceles, scalene, right-angled	A triangle is a polygon with three sides.

equilateral   isosceles

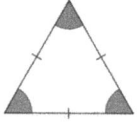 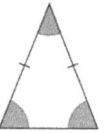

three equal sides   two equal sides

scalene   right-angled

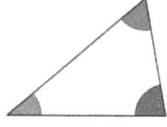 

no equal sides   one angle is 90°

two-dimensional (2D)	A flat shape has two dimensions, length and width or base and height.
unknown	An unknown is a variable. You can often find its value by solving an equation.
value	The value is the amount an expression or variable is worth.
variable	A variable is a symbol that can take any value.
vertex, vertices	A vertex of a shape is a point at which two or more edges meet.

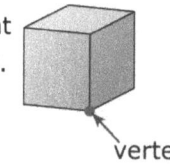

vertex

vertical	Vertical means straight up and down.
vertically opposite angles	When two straight lines cross they form two pairs of equal angles called vertically opposite angles. 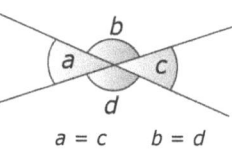 $a = c$ $b = d$
whole	The whole is the full amount.
width	Width is a dimension of an object describing how wide it is.
x-axis, y-axis	On a coordinate grid, the x-axis is usually the horizontal axis and the y-axis is usually the vertical axis. 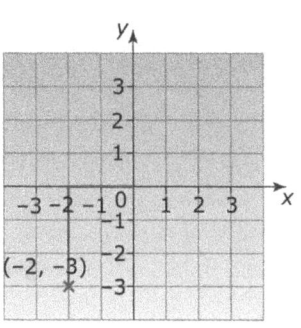
x-coordinate, y-coordinate	The x-coordinate is the distance along the x-axis. The y-coordinate is the distance along the y-axis.
	For example, (−2, −3) is −2 along the x-axis and −3 along the y-axis.
zero	Zero is nought or nothing. A zero place holder is used to show the place value of other digits in a number. 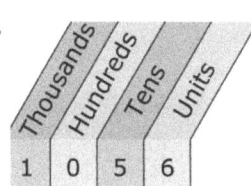